SpringerBriefs in Philosophy

More information about this series at http://www.springer.com/series/10082

Zuzana Parusniková

David Hume, Sceptic

Springer

Zuzana Parusniková
The Czech Academy of Sciences
Institute of Philosophy
Prague, Czech Republic

ISSN 2211-4548 ISSN 2211-4556 (electronic)
SpringerBriefs in Philosophy
ISBN 978-3-319-43792-7 ISBN 978-3-319-43794-1 (eBook)
DOI 10.1007/978-3-319-43794-1

Library of Congress Control Number: 2016953341

Printed on acid-free paper

This Springer imprint is published by Springer Nature
The registered company is Springer International Publishing AG Switzerland

For Robert

Preface

Hume wrote his philosophy in a time of tumult. The millennia-old metaphysical tradition that placed humans and their cognitive abilities in an ontological framework collapsed and gave way to a new one that placed the autonomy of the individual in its centre. Subsequently, the world and life lost their deeper meaning, and nature was reduced to a collection of objects for reason and science to explain and master. It was the birth of modernity that Descartes inaugurated and Kant completed with his Copernican revolution. Modernity gave rise to a new kind of scepticism, involving doubt not just about the adequacy of our knowledge but about the very existence of a world independent of the self. Firmly placed within this philosophical framework, Hume faced this scepticism implying phenomenalism. His empiricism added yet another sceptical theme: how can one, on an empirical basis, rationally justify key concepts of human understanding, primarily the ideas of causation and distinct existence, produced instinctively by the human mind? Moreover, Hume was influenced by Pyrrhonian scepticism, much discussed by philosophers at the time. The Pyrrhonian legacy is especially noticeable in his acceptance of the weakness of reason and in his emphasis on the practical role of philosophy. Rather than serving as the foundation of science, in his hand philosophy became a guide to a joyful, happy life, to moderately critical observations of common life and to an active involvement in society. This way Hume's philosophy stood in strong opposition to the (early) modern mainstream.

Keywords David Hume • Scepticism • Epistemology • Science of man • Cartesian paradigm • Pyrrhonian crisis • Philosophy of life

Contents

List of Abbreviations

T Hume, D. [1739]. *A Treatise of Human Nature*. Ed. Selby-Bigge, L.A. Revised Nidditch, P.H. Second edition. Oxford: Clarendon Press. 1981.

E Hume, D. [1748]. *An Enquiry Concerning Human Understanding*. Ed. Selby-Bigge, L.A. Revised Nidditch, P.H. Third edition. Oxford: Clarendon Press. 2005.

DL Diogenes Laertius [c. 300 AD]: *Lives of Eminent Philosophers*. Volume II. Trans. Hicks, R.D. London & New York: G.P. Putnam's sons. 1925.
(*DL* IX, 78, 491) – Book, Chapter, page

PH Sextus Empiricus [c. 200 AD]: *Outlines of Pyrrhonism* (*Pyrhoneioi Hypotypoesis*). Eds. Annas, J. and J. Barnes. Cambridge: Cambridge University Press. 2000.
(*PH* I, 13, 6) – Book, Chapter, page

Met. Aristotle 1993 [c. 335–323 BC]. *Metaphysics*. In *Aristotle in 23 Volumes*,. Volumes 17, 18. Trans. *H.Tredennick.* London: William Heinemann Ltd. 1933.
(*Met* XII, 7, 1072b) – Book, Chapter, column
Pagination: August Immanuel Bekker *Aristotelis Opera edidit Academia Regia Borussica*, Berlin, 1831–1870.

DA Aristotle 1903 [c. 350 BC]. *De anima*. Trans. R.D. Hicks. Cambridge: At The University Press. 1903.
(*DA* III, 5,429b) – Book, Chapter, column
Pagination: August Immanuel Bekker *Aristotelis Opera edidit Academia Regia Borussica*, Berlin, 1831–1870.

Introduction

Hume's epistemology has been a subject of controversy for almost 300 years. What has made it unsettling but also exciting for philosophers of many generations is undoubtedly its sceptical slant. Hume demonstrated the severe limitations of our reason and knowledge; reason can be described as a leaky, weather-beaten vessel rolling on the waves of doubt. Hume's scepticism came under much criticism from his contemporaries. He was writing at a time when epistemology was coming to the forefront of philosophical interest – traditional scholasticism was in crisis and new scientific discoveries, independent of the Aristotelian scholastic worldview, offered new explanations of the workings of nature and the universe. New epistemological foundations for science were needed, and Hume's inquiries into the operations of the mind, based on experience and observation, were a part of these efforts. His sceptical findings, however, undermined his goal. Hume was not alone in tackling the dangers of scepticism. The revival of Pyrrhonian scepticism played a formative role in the development of early modern epistemology, and both philosophers and scientists tried to eliminate or at least to minimize its negative impact on the status of knowledge. This is also a reason why Hume's scepticism has attracted such great interest.

In the twentieth century, Hume's grand project of the science of man – comprising our reasoning, feelings and sentiments and our behaviour and social life – began to be more appreciated, and epistemological scepticism was often reconsidered in this broad context of his writings. Though epistemological scepticism is only a part of the whole system of sciences about human nature, it does deserve attention as such; moral philosophy, though more rewarding for a Humean philosopher, does not change much the key sceptical contradictions concerning human understanding. In this book, I examine Hume's *epistemological* scepticism in a broader context. In order to do that, three aspects of Hume's scepticism should be mentioned, namely, the Cartesian, the empiricist and the Pyrrhonian. The first shows Hume's commitment to the new philosophical paradigm based on the autonomy of the individual mind; the second exposes the limitations of an epistemology based on experience; the third explains Hume's commitment to the ancient view that scepticism should cultivate modesty and tolerance and guide us to happiness.

The *Cartesian strand* of Hume's scepticism unfolds from the revolutionary change in the seventeenth-century philosophy that can be called a *turn to the subject*; the old metaphysical framework in which everything, including our mind, was structured by a universal order was replaced by the independent human mind that broke free from any higher authority and became an autonomous and constructive cognitive agent. Subsequently, the old ontological self-evidence of the world came under sceptical attack. The autonomy of the mind liberates man from metaphysical bonds but, at the same time, imprisons him within his subjectivity. How can we establish any existence beyond our mental world and bridge the gap? It seemed at the time that this mystery could not be resolved without the use of metaphysics, the speculative and abstruse philosophy to which Hume would never resort. But then, the assumption that all we can know are the contents of our minds leads to phenomenalism. Descartes was the first to advance this argument but withdrew it after realizing that consistent phenomenalism would disable science by taking away its very subject – the external world – and by limiting knowledge to subjective certainty about phenomena.

In this respect, *Hume was the most consistent adherent to the early modern epistemological discourse*. The problem of where the contents of our mind come from is for him beyond the narrow compass of human understanding; and speculations of any kind do not belong, he believed, to experimental philosophy. Since epistemology does not concern the external world, Hume shifts its focus to internal mental processes. It is concerned with mapping the mind itself, investigating its operations and capacities. Hume redefined a new area of philosophical inquiry that led to the emergence of disciplines like the philosophy of mind and cognitive science. Yet these inquiries brought to light another level of scepticism that has its roots in Hume's empiricism.

The *empiricist strand* of Hume's scepticism targets the mechanisms of the mind by which we tie together impressions and construct ideas. Empiricism proceeds from the rule that all mental procedures – in order to qualify as rationally legitimate – must be based on observation and experience. This rather drastically limits the scope of epistemology: ideas must be derived from impressions and correspond to them. Hume showed that the fundamental type of inference that we use in our reasoning – causation – breaks this rule by assuming that two events we observe regularly conjoined are necessarily connected. Likewise, we presume that impressions experienced conjoined at some point have distinct existence. But neither necessary connection nor distinct existence can be derived from experience. Hume's conclusion seems damaging for the status of our knowledge and most importantly of science; and in Hume's time of advancing modernity, culminating in the Enlightenment endorsement of reason and its power to master nature through knowledge, such a conclusion was very disturbing indeed. It was a time of great scientific surge, accompanied by great epistemological confidence and optimism. Epistemology was supposed to provide rational foundations for science, and this was what Hume could not deliver. Curiously, Hume scholars both past and present seem to be much more worried about this than Hume ever was.

An explanation is possible once we situate Hume in a broader sceptical context. A *strong influence of Pyrrhonism on Hume's thought* explains why he so easily accepted the weakness of reason as a predicament we have to live with. In accord with Pyrrhonism, this need not be devastating, as most modern epistemologically oriented philosophers are disposed to think. Hume's corrected (for some, perfected) version of Pyrrhonism follows on the old sceptical principle that we tranquilly stick to appearances but newly asserts the power of natural inclinations in life. Hume's Pyrrhonism provides an intriguing perspective on the relation between our reason and our nature; instead of attempting to find some strong *arguments* against scepticism, Hume daringly asserted a disparity between them, yet without any pessimism.

This is largely due to the Pyrrhonian influence with regard to the purpose of philosophy. Instead of endeavouring to be the foundation of science, it has its domain in the practical and moral spheres; it is concerned with *life* and with living with others. Philosophy becomes a guide to a balanced, happy life (Hume's version of *ataraxia* spiced by earthly pleasures), advocating modesty and humility as opposed to dogmatism and fanaticism, a respect for tradition and customs and active involvement in public affairs. Having served as the source of the most lethal epistemological scepticism, Pyrrhonism now helps shift Hume's interest from epistemological themes to moral philosophy. *Hume thus stands outside the dominant epistemological discourse of (early) modern philosophy*. His position can be seen as almost scandalous – waving away 'burdensome' scepticism in epistemology and calmly moving to other areas that provide more philosophical satisfaction. Various aspects of the ancient Pyrrhonian strand of Hume's scepticism – still waiting for an adequate appreciation – will be presented as crucial for a thorough understanding of his philosophy.

All three sceptical strands of Hume's scepticism – Cartesian, empiricist and Pyrrhonian – create a multilayered picture of his epistemology and introduce him as a fascinating figure standing between the ancients and the moderns, taking each position to its logical conclusion. Hume was influenced by Pyrrhonism to a greater degree than were his fellow philosophers and in the more original sense that ascribed philosophy a practical therapeutic role in life. But he also stood in the forefront of the modern discourse by recognizing the implications of the Cartesian turn to the subject and, subsequently, by a consistent acceptance of phenomenalism; the existence of external reality is not denied, but it cannot be established either. By defining the purpose of epistemology as an inquiry about the internal world of our minds, Hume anticipated many modern philosophical trends.

Chapter 1
The Cartesian Roots of Hume's Scepticism

Abstract The Cartesian strand of Hume's scepticism unfolds from the revolutionary change in the seventeenth century philosophy that can be called a *turn to the subject*; the old metaphysical framework in which everything, including our mind, was structured by a universal order was replaced by the independent human mind that broke free from any higher authority and became an autonomous and constructive cognitive agent. Descartes was the first to advance this idea and caused a paradigmatic change in philosophy that opened the modern era. Hume developed his philosophy on this ground and the fact that Hume turns to experience while Descartes turns to pure intellectual insight testifies to a secondary level of differences. The autonomy of the mind liberated man from metaphysical bonds but, at the same time, imprisoned him within his subjectivity, leading to phenomenalism. The question "what is beyond our mind?" becomes meaningless for Hume – it is something beyond experience and we have to suspend our judgment on this issue. Our beliefs and imagination are of course unaffected by this scepticism but are beyond the jurisdiction of reason.

Keywords Metaphysics • Being • Mind • Modernity • Epistemology

1.1 The Rise of Early Modern Philosophy

The greatest change in philosophy ever is arguably the seventeenth century shift from scholastic to modern philosophy associated with Descartes.[1] He turned everything upside down: our perception of the world, the place of man in it, the purpose of cognition, the status of mind, key values – the whole philosophical worldview, in fact. Descartes announced a new philosophical start free of the metaphysical burdens of the past. He brushed aside the old wisdom that formed the basis of the cultural, intellectual and religious climate in the preceding millennia, and set a new terrain for

[1] For simplicity, I shall use the term 'modern philosophy' instead of 'early modern philosophy'.

Z. Parusniková, *David Hume, Sceptic*, SpringerBriefs in Philosophy,
DOI 10.1007/978-3-319-43794-1_1

philosophy founded on the autonomy of the self.[2] Descartes's historical significance was in his revolt against any authority that transcends man and his individual reason. Man as the highest rational authority acquired unprecedented privileges and philosophy shifted its focus to studying human cognitive capacities and limits, with the aim of building a new science of nature.[3] Descartes' paradigmatic change in philosophy can be compared to the revolution in science associated with Newton.

Hume's philosophy grew from this foundation though Hume endeavoured to create a new science of man, not of nature. But he took it for granted that philosophy starts with man and his mind, not with Being and its ontological structure. Hume considered the mind and its operations a primary focus of philosophy and the basis of other sciences. In this sense epistemology was a foundational discipline, though in the end Hume had to accept defeat regarding the rational foundations of knowledge. The fact that Hume turns to experience while Descartes turns to pure intellectual insight testifies to a secondary level of differences. The new focus of philosophy, the status of the mind, and the rejection of metaphysics represent their common ground. The problem of establishing access to – or the mere existence of – the world was a new challenge to be faced, after the world and our knowledge about it lost the unproblematic self-evidence that it had in the age of metaphysics. These issues constitute the deepest form of scepticism and link Hume to Descartes. Hume understood these implications of the position of the autonomous self perfectly, unlike some of his fellow-empiricists (e.g. Gassendi or Locke).

To understand the paradigm-change triggered by Descartes one has to view it in the context of the prevailing metaphysical grounding of philosophy at that time. Though somewhat eroded, the Aristotelian-scholastic tradition represented the official doctrine that formed the rules and set the limits for both philosophy and natural science. These were the teachings of the schools, speculative and abstruse, that Hume opposed so strongly; the path to building the new philosophy, based on the freedom of thought, was already paved for him by Descartes. Flew notes that Descartes "suddenly releases a shattering salvo of almost all-destroying doubt", referring to his statement in the *Discourse* "I reject as false all the reasons formerly accepted by me as demonstrations", followed by the rock-solid certainty "*I think, therefore I am*", and claims that "the modern period in philosophy starts with this devastating sentence" (Flew 1986, 12–3).

Hume developed his philosophy on this new philosophical scene. Flew identifies three interlocking Cartesian elements that influenced Hume (and more generally, the British empiricists), of which one is crucially important: "we are never immediately and non-inferentially aware of anything outside and independent of ourselves" (Flew 1986, 15). This leads to phenomenalism, a new and acute problem for modern

[2] I am referring to Descartes' revolutionary step of *shifting the focus of philosophy from Being to the Self*, as spelled out partly in the *Rules*, in his *Discourse*, and with most philosophical precision in the first two *Meditations* (and the first half of the third). For the time being I do not include Descartes' restoration of metaphysics, a step that he took in order to solve the problem of the independent existence of the world and the possibility of true knowledge.

[3] I shall use the male form "man" throughout to stay in line with the texts of Hume's time.

philosophy. Norton shares this opinion and with reference to the influence of Descartes, Locke, Bayle and Malebranche on Hume he remarks: "Hume was satisfied that the battle to establish reliable links between thought and reality had been fought *and lost*" (Norton 2009, 12). The fact that Hume takes perceptions – whose source is unknown – to be the only immediate objects of mind caused, according to Norton, a phenomenological turn in Hume's epistemology.

The basic Cartesian principles seem to be so firmly embedded in the present analytically oriented discourse: the focus on the mind, the link between philosophy and science and the problem of an independent status of objective reality are still alive in current debates between realists, externalists, instrumentalists, contextualists and projectivists. The fact that the Cartesian legacy is still alive and taken for granted can obscure how enormously different it was to pre-modern philosophy. A brief detour into the pre-Cartesian metaphysics can help to elucidate this difference.

1.2 Metaphysics

To understand how the shift from metaphysics to the autonomous cognizing mind broke the ties with the past can be best explained by contrasting some central themes in metaphysics with the position of modern philosophy I shall focus on two crucial differences between metaphysics and the newly emerging philosophy to illustrate the incommensurability between them.

Despite huge differences between various ancient and scholastic conceptions, two beliefs were formative in the metaphysical tradition. First was that a *metaphysical grounding* united all being in a meaningful whole, comprising of heavens, nature and man. There was a hierarchy of various degrees of being that reflected the degree of participation in the divine Being. Second, man, due to his reason and knowledge, was a privileged being capable of the highest degree of participation. Yet *cognition was integrated in the all-encompassing order and was subordinate to it*. Cognition was an ontologically anchored process integrated within the structure of Being.

In the ancient perspective, the metaphysical structure of Being incorporated the universe, our world and our life; the ordered structure of the universe (*Lógos*) was a barrier to chaos whose monstrosity is hardly describable in words; an amorphous swamp with no shape or limits, darkness and emptiness, nothingness.[4] The ancient metaphysics introduced the concept of Logos that is the ordering principle of the cosmos (it *is* cosmos) and makes it intelligible to us. This 'disciplining' meaning of

[4] A metaphoric description can be found in Hesiod's *Theogony*: "And there, all in their order, are the sources and ends of gloomy earth and misty Tartarus and the unfruitful sea and starry heaven, loathsome and dank, which even the gods abhor. It is a great gulf, and if once a man were within the gates, he would not reach the floor until a whole year had reached its end, but cruel blast upon blast would carry him this way and that. And this marvel is awful even to the deathless gods." (Hesiod 1914, lines 736–744).

the term '*Logos*' is limit or boundary (*peras*) with an almost combative function in keeping the universe safe from chaos; Homer uses the word to describe an ordered rank of soldiers ready for battle. The emphasis on the rational structure of *Logos* can be found in the Pythagorean and later in the Platonic conceptions of numbers, geometrical shapes and tones conceived as *ontological entities* organized in a harmonious whole. The universe is structured in perfect mathematical proportions and numerical ratios have a counterpart in harmonious sound frequencies; sounds linked to certain numbers create an orbital resonance in the universe; the movement of celestial bodies is in consonance with their distance from the Earth.

These Pythagorean ideas influenced Plato (and later, in the Renaissance, a number of philosophers and scientists like Robert Fludd and Johannes Kepler). The mathematical structure of the universe was addressed by Plato in his *Timaeus*. He identifies four basic building elements of the Earth that have the form of geometrical entities: earth is a cube, fire is a pyramid, water is an icosahedron and air is an octahedron; these entities are the embodiment of perfection since they are composed of regular polygons (triangles, squares and pentagons).[5] In this conception the world has precise numerical proportions, linked to the harmony of tones – a legacy of Pythagoras. The universe, too, is constructed according to the rational model; the Earth, sun and stars in the shape of round spheres and the soul as two concentric circles with one center. Unlike in Christianity, Greek philosophers did not try to pinpoint the ultimate origin or source of the universe; it was assumed to be eternal, with its own inherent structure, dynamics and interrelations.

Plato's conception seems to encourage a mathematical analysis of nature; it inspired, for instance, Galilei and Kepler. Nature that has a mathematical structure and can be dissected into parts and expressed in an idealized form – in equations and graphs – seems ideal for modern science. Could modern science not have developed the Platonic model to suit its needs? This question takes us to the original problem concerning the incommensurability between the two philosophical discourses. For Plato, knowledge has a metaphysical anchor and a wider purpose. The goal of knowledge is the search for the ideal essences, the Forms, that are the highest entities in the universe. Forms are present in our souls but are buried under the silt of the bodily burden that taints their purity. And since Forms are eternal our cognition is not a discovery of something new but a recollection of the forgotten, moving us beyond this world towards the pure light of Ideas.

By contrast to this conception, modern epistemology is charged with enthusiasm to discover new facts and truths, ready for action. The spirit of modern philosophy is active and adventurous, mobilizing man to use his reason to explore the world anew, achieving progress in knowledge due to his own abilities. For Plato, philosophical knowledge is rather contemplation on the divine that brings the ultimate joy (*eudaimonia*). Philosophers are able to get into the proximity of Ideas when they contemplate the highest virtues and values, like beauty, justice, truth, love or numbers; philosophers are thus able to 'touch' the highest idea of Good. In this picture,

[5] The inscription above the entrance to Plato's Academy in Athens said: "Let no one ignorant of mathematics [geometry] enter this door".

our souls have in themselves an intermediary, a spiritual force, to help them to approach the essences; it is a *daimon*, daemon, Socrates' *daimonion*. Thus, knowledge is not an indifferent – cold and analytical – business that itself represents the highest form of inquiry, but contains aesthetical and moral dimensions. The Platonic approach does not aim at conquering nature through knowledge, but rather appreciates the allure of the geometrical and musical harmony of the cosmos, its beauty and elegance. Like *Logos* the term 'cosmos' has a wide range of meanings, one of which denotes the perfection of the whole; the Greek verb '*kosmein*' refers to having or maintaining perfect order, with a slight exaggeration 'to adorn' – from which comes the word 'cosmetics'.

1.3 *Nous*

From some viewpoints, Aristotle's conception of knowledge might seem particularly science-friendly. Aristotle established a wide range of scientific disciplines such as astronomy and cosmology (with an emphasis on the movement of heavenly bodies), natural science including geography, meteorology, geology, physics, biology and the study of psychic and cognitive processes; logic; social sciences like politics and ethics, and art. His system of sciences formed the pillar of scholastic natural philosophy. Aristotle valued experience as the starting-point of knowledge and argued that sensory perceptions are not deceptive; even in sensory knowledge truth displays itself, albeit partially, and rational activity must follow to capture the *eidos* (the pure form, essence) of the thing under investigation. Could modern science not have drawn on the Aristotelian legacy by extending or correcting it?

The reality was quite the opposite; Aristotle's system was even less suitable for modern science than Plato's which was better fitted for the ideal of mathematical science. Aristotle, in contrast, presented a *qualitative conception of nature* in which each entity – in the realization of its general *telos* as a species – undergoes a unique process with individual variations. For instance a tree fulfils its purpose by the process (motion) of growing, blossoming and thus providing pollen for bees, bearing fruits, dispersing seeds; tree as a species has as one ideal form in common yet in the realization of this form there are individual irregularities in concrete execution, either due to circumstances or to individual disposition. As soon as matter (*hýlé*) becomes formed by the purposeful movement of bringing some *eidos* to actuality – and we never encounter pure formless matter; the undefined matter exists nowhere – it enters the process of qualitative transformation. Modern science, by contrast, needed the conception of uniform matter without any internal purpose, matter that can be decomposed to its elementary components that can further be quantified; e.g. atoms (proposed in modern terms by Gassendi). Modern scientists had to cut themselves loose from the then dominant Aristotelian system.

The qualitative motions in nature are part of the ontological processes conforming to universal teleology. Individual movements are interlinked; they form and sustain the whole into which they belong. In Aristotle's words, "all things are ordered together somehow, but not all alike, for example fishes and fowls and plants; and the world is not such that one thing has nothing to do with another, for all are ordered together to one end (*pros hen*)" (*Met.* XII, 10, 1075a). Human cognition belongs to this process, too; it is conceived by Aristotle as one movement among many, though privileged due to its rational component. A cognitive act is performed by an individual but has also an ontological dimension. This dimension must be approached from the perspective of the highest being – the divine intelligence, God, *Nous*, thinking that thinks itself *(noésis noéseós)*, being that is eternal, supremely perfect and unmoved (unmoving).

As a perfect being *Nous* cannot be in motion since motion indicates imperfection and thus is contrary to *Nous*. The only movement occurs within it, *Nous* thinks itself, which is an inward, self-reflexive movement since thought (reason) is the fundamental ontological principle. *Nous* is entirely self-contained and, metaphorically, it relishes its own being and has no wish to be disturbed. But Aristotle defines *Nous* as the unmoved first mover (*to protón kinún akinéton*); this definition evokes the picture of some first "push" by which God sets in motion the universe including our world, or of the act of divine creation of the world, as in Christianity. This picture is mistaken: outward activity would violate the perfection of *Nous*. It is the other way round. *Nous* causes motion by waking the desire in the lower spheres of being to reach the same or the maximum possible perfection and get to the closest proximity of *Nous*. In Aristotle's words, *Nous* "causes motion as being an object of love, whereas all other things cause motion because they are themselves in motion" (*Met.* XII, 7, 1072b).

The eighth superlunary sphere of stars whose perfection is just below *Nous* desires to reach the perfection of *Nous*, and begins to move in the most perfect way – in a circular and uniform motion; in this sense only is the *Nous* the first mover. This motion is then transferred downwards to other celestial spheres, the Sun and the planets, in complicated ways; for instance, each planet has four spheres and the planet itself is placed in the fourth one. There may be a hint in Aristotle's account that separate lower-rank movers are responsible for the motion of the lower-rank celestial bodies. This motion extends to the sublunary sphere into our world, the world of changing (moving) Nature. Our cognition, too, belongs to this pattern. All qualitative movements in the universe are motivated by the *internal desire* for improvement that is present in each kind of being – they have their own internal dynamics. Each thing is compelled to reach the highest possible perfection in its rank by realizing its form, *eidos*, by bringing its potentiality to actuality and, on the higher level, to contribute to and to co-create the perfect whole.

This illustrates the *ontological and qualitative dimension of our cognition* and in a broader sense the dependency of epistemology on ontology. On the one hand, cognition is a subjective mental act; a movement that is driven by the internal desire of the soul for perfection, as expressed in the famous statement from the beginning of Aristotle's *Metaphysics*, "All men by nature desire to know" (*Met.* I, 1, 980a). In knowledge, the potential of the soul is realized by grasping the *eidos* of the object of cognition; "the mind is the power of becoming such objects without their matter" (*DA* III, 4, 429b). Human cognition surpasses all other movements by the intellectual surge to truth and by the ability to think the divine; that privilege is reserved for philosophers whose intellect is best suited for the task, and only for certain moments: "the life which he [*Nous*] lives [blissful self-contemplation] is the most excellent, and such that we enjoy for a small portion of time; such a life is with him perpetual" (*Met.* XII, 7, 1072b). Therefore, man represents a higher degree of being than, for instance, a tree. On the other hand, everything moves *by its own nature* in sustaining the cohesive whole of the universe, and our cognition, too, belongs to this ontological movement.

Human understanding unfolds on two level – the soul (senses and reason) fulfills its own purpose by grasping the *eidos* of the thing, and inversely, the *thing reaches fulfillment in this act of being grasped*; this is an interesting point since a thing exists in fact twice – first in reality when it realizes its purpose as a thing, and second as the immaterial *eidos* that is realized in our understanding. This further manifests the ontological interconnection between nature, cognition and the universe. Cognition, by grasping the ideal form of things, ties the soul (mind) and the world together. This ensures that "the actual knowledge is identical with the thing known" (*DA* III, 7, 431), that reason, when it thinks, is the same with objects without their matter (grasps the *eidos* of objects). The same ontological founding of the universe, nature and the mind guarantees not only the real existence of nature (the question about the ontological status of nature could have never arisen in metaphysics) but also the possibility of true knowledge – the correspondence between our thoughts and objects.

Knowledge that exists only potentially is for Aristotle *nothing* and, accordingly, our reason is nothing before it begins to cognize (to move); reason becomes real only when the cognitive process itself begins. As Aristotle observes "the mind [soul] is in a manner potentially all objects of thought but it is actually nothing until it thinks: potentially in the same sense as in a tablet which has nothing actually written upon it. This is exactly the case with the mind … the mind is the power of becoming such objects without their matter" (*DA* III, 5, 429b). The description of the mind as an empty tablet, "tabula rasa", was used by Locke (in his reference to "white paper"; no wax tablets were used in his time) in his concept of the mind that receives its content from the senses (Locke 2011, 104). But this is not what Aristotle meant in this example. He made an analogy between the mere potentiality (nothingness, empty tablet) of both reason and sensation on the one hand, and motion (realization of purpose, *energeiá*) of both reason and sensation on the other, while the motion springs from the soul itself, from its internal tendency to realize its potential. Aristotle does not in this particular passage refer to the priority of experience but makes a distinction between the (empty) potentiality and the actuality (knowledge). Hegel correctly interprets this passage from Aristotle: "understanding of a surety … has not the passivity of a writing tablet; it is itself the energy, which is not, as it would be in the case of a tablet, external to it … The soul is this book unwritten on, a book that contains all things potentially, but in reality contains nothing before it is written on. Before real activity nothing truly exists" (Hegel 1894, 196–7).

This brief excursion into metaphysics demonstrates that despite many important differences between the individual systems at least two principles were shared –

conceiving the universe and nature as a meaningful whole, and viewing our cognizing activity as integrated within this whole; *our mind does not stand on its own*, but has a deeper grounding and purpose. Even the activity of our soul (mind, perceptions and reason) contributes to the self-structuring processes in the universe. This approach is alien to the modern spirit that takes knowledge, ideally scientific knowledge, to be a value-indifferent scrutiny of external value-free objects. In modernity, objects in Nature are made of universal matter and have no internal, intimate relation to the individual mind. The metaphysical approach existed – in many variations – from the ancient and medieval times. Descartes broke this millennia-long tradition in a simple announcement: "My plan has never gone further (*sic!*) than an attempt to reform my own thoughts and rebuild them *on ground that is altogether my own*." [italics mine] (Descartes 1979a, 18).

1.4 The New Start

Due to Descartes the metaphysical scheme in which man had its secure place began to disintegrate under the pressure of the confident Self, enforcing its rights to be the ultimate authority and the judge as to what can be known and how we should proceed in acquiring knowledge; the Self became the starting point, and knowledge the primary focus, of philosophical investigations. In the previous metaphysical scheme, *man was at home in the world* and cognition, although a privilege, was not 'altogether his own' but was a part of other movements and processes in nature. Man could let himself be carried along by the cognitive process in which things naturally revealed themselves to him; Hume later ascribes this role to nature but for Aristotle, the 'current' had a metaphysical founding. The order of being corresponded to the order of thinking and thus there was an unproblematic 'match' between the two, provided we use our senses and reason appropriately; our cognition was a part of the universal teleology. Now, Descartes challenged us to "get rid of *all* our previous wisdom, renounce *all* our opinions, and make ourselves free of *all* blindly accepted certainties, to reject *all* existing authorities" (Koyré 1979, xx).

It may seem from this account as though Descartes appeared like a bolt of lightening out of the blue, and demolished all what had existed before. But there had been signals of crisis of the scholastic system before Descartes. Most important was the Renaissance change of focus from the heavens to the earth, to our human dimension, to life and our intellectual capacities. Man began looking at the world anew, with a childlike curiosity, being enchanted by the new space, the beauty and the mysteries to be discovered. New vision in the arts, new considerations of space, new experiments in physics, new technologies, alongside with the discoveries of new continents – these all ignited the desire to break the shell of the Aristotelian-Ptolemaic finite cosmos and to break free from the official scholastic doctrines concerning science. In this tumult man claimed the right to think without the straightjacket of the scholastic picture of the universe, and the resulting tensions could not be reconciled. The time was ripe for a change. The revival of scepticism in Europe played an important role in deepening the crisis of scholastics (see Chaps. 4 and 5). The sceptics claimed that any judgment is contradictory in itself and thus, under the sceptical scrutiny, disintegrates from within. This sceptical principle applied for the scholastic arguments for the God's existence and thus undermined the very foundation of theology.

But these symptoms of a crisis of scholastics cannot diminish the philosophical breakthrough sparked off by Descartes; he was responsible for the emergence of a new paradigm that influenced philosophy development for centuries and is influential even now. Though Descartes was educated at the top-rank Jesuit College at La Fleche in the metaphysical tradition, the old wisdom did not impress him. After his studies at the College Descartes set out to study the world; he travelled several years and observed various ways of life. Eventually he retreated to his chamber, sat by the fire and carried out intensive reflections on philosophy. He condemned the whole metaphysical tradition as a relic and announced a new start. This new philosophy formed, in most general terms, the basis for modern philosophers including Hume; Hume was the most radical in accepting the new principles, entailing the liberation of man from the metaphysical burden and putting him – and his reason – to the fore. Cognition was conceived as a purely individual human activity – which may, of course, seem obvious today but in Descartes' time was unheard of.

Already in his first work, *The Rules for the direction of the mind*, Descartes proclaimed already in that "we are now freed from the oath which bounds us to our master and are old enough to be no longer subject to the rod. So if we seriously wish to propose rules for ourselves which will help us scale the heights of human knowledge … we should take care not to waste our time … by occupying ourselves only with difficult matters"; the master, of course, being Aristotle (Descartes 1985, 11–12). The manner of philosophizing practiced in the schools, he said, could only serve as mental exercise for students to refine their abilities of argumentation (Descartes probably meant the Jesuit art of dialectics). Further in the *Rules* Descartes blamed the old scholastic system, especially Aristotelian teachings about nature, for being foolish – for having grand ambitions to study the secrets of nature *before analyzing the powers of reason first*; it is, of course, unfair to blame Aristotle for this but Descartes had already a different concept of reason in mind.

Descartes grounded all cognitive powers in man alone. The new primary task for philosophy was to determine the norms of the conduct of the mind in order to achieve certain knowledge. The important principle in these efforts was that we are our only guides and judges. Once we investigate our cognitive faculties we can define the right method for science; knowledge can then be achieved easily, *as if by play*. This feature of modern philosophy anticipates the 'democratic' spirit of the Enlightenment: reason and proper judgment are accessible to everybody, even to a labourer or a clerk, since we all have the same rational potential, the same cognitive dispositions. Knowledge is not reserved just for privileged scholars who spend years in schools studying old books and pursuing endless learned disputes. The new method makes it possible for knowledge to advance swiftly; Descartes believed that mastering nature through knowledge would be a matter of a few decades! It was to be a new start for man and a new dawn for science. We can almost feel Descartes' eagerness and his impatience to get to work and rebuild the whole of human knowledge.

Hume chose a different method and a different goal; instead of natural science he focused on the science of human nature, thus widening the scope of philosophy far beyond epistemology. And instead of the light of reason he chose experience and observation as the pillars of epistemology. But he displays similar eagerness and determination to do away with metaphysics and start from a completely new foundation, from the human mind. Hume, too, abandons the "tedious lingering method, which we have hitherto followed … [since] this obscurity in the profound and abstract philosophy … is not only painful and fatiguing but is the inevitable source of uncertainty and error" (E 11–12).[6] Accordingly, in the 'Introduction' to his *Treatise* Hume set out to "propose a compleat new system of the sciences, built on a foundation almost entirely new and the only one upon which they can stand on with any security" (T xvi). The only foundation to be built on is one that man alone will choose without interference from any higher power. But for both Descartes and Hume, then, *philosophy starts from the human mind and its cognitive capacities*. Discovering the powers and limits of the mind, and defining the proper method of judgment are the first steps in building the new system of knowledge; "accurate and just reasoning is the only catholic remedy, fitted for all persons and all dispositions; and is alone able to subvert that abstruse philosophy and metaphysical jargon…" (E 12). And Hume, too, cannot wait "to enter upon the enterprize with thorough care and attention" (E 15). The second, trickier step was to establish the extra-mental reality.

Descartes clearly stipulated that only such knowledge that has undergone *man's individual scrutiny* may be accepted: "the first [rule of logic] was never to accept anything as true if I had not evident knowledge of its being so; that is, carefully to avoid precipitancy and prejudice, and to embrace in my judgement only what presented itself to my mind so clearly and distinctly that I had no occasion to doubt … [only] what squared with the norm of [my] reason" (Descartes 1979a, 17–20). In his *Abstract* Hume argued in a similar way: "*the sole end of* logic *is to explain the principles and operations of our reasoning faculty*" (T *Abstract* 646). Epistemology had a wider meaning for Hume than for Descartes – it included psychological investigations into the human mind. But both philosophers were determined to build philosophy on the principle of the autonomy and authority of the individual.

Though Hume wrote a century after Descartes he still felt the negative impact of metaphysics; it was for him an example of bad, dogmatic and oppressive teaching. Descartes witnessed the fate of Giordano Bruno which scared him so much that he put on hold the publication of one of his books (*The World*). A charge of heresy might have endangered his life. While this would not have been possible in Hume's time, the case of a young student from Edinburgh, Thomas Aikenhead, executed at the age of 20 for blasphemy as late as in 1697, remained in memory; his age was no excuse, and nor was the fact that he mocked the Bible while drinking with friends in a pub. That was 14 years before Hume's birth. But Hume protested less against

[6] This may explain Hume's view that Aristotle is "utterly decayed" while Cicero flourishes (the praise of Cicero is understandable in the context of the renewed interest in scepticism in that time; however, Hume's conviction that Locke will soon be forgotten is not easy to understand). See E 7.

subtleties in the scholastic teachings than against a more general vice. He blamed metaphysics for tempting people to pursue subjects that are not accessible to human understanding. Hume also rejected the political power of the Church and its institutional control over free thought. He himself experienced its impact when he was twice rejected for university posts on the charge of atheism. Hume warned against all kinds of metaphysicians and compared them to robbers who "lie in wait to break upon every unguarded avenue of the mind to overwhelm it with religious fears and prejudices' and even the 'stoutest antagonist, if he remit his watch a moment, is oppressed" (E 11).

Both philosophers believed in taking radical action against the old wisdom. As Hume says, "instead of taking now and then a castle or village on the frontier, [we must] march up directly to the capital or centre of these sciences, to human nature itself" (T xvi). Descartes shows no mercy to the old knowledge either. After spending a day alone in his stove-heated room he decided that metaphysics cannot be reformed because it stands on bad foundations that have been patched together from incongruous parts over time. He uses an architectural metaphor: "the ancient cities … are as a rule badly laid out … streets are twisted and irregular … as compared with those towns of regular pattern that are laid out by a single designer on an open plain towns; … but when the foundation is undermined, the superstructure will collapse of itself" (Descartes 1979a, 15). He proposes a geometrical plan of the town (knowledge), with wide straight boulevards designed by reason. Hume turned to a different architect – to experience – but with the same intention of redefining the system of knowledge. It is interesting that Hume appeals to the respect for tradition in the moral and social spheres, but in epistemology he preferred a radical revolt against the old. For both Descartes and Hume, the old buildings that represent the previous metaphysical knowledge are now collapsing because they rest on rotten foundations. What remains in place of the debris? Man standing proudly alone, ready to design an entirely new plan of knowledge resting on the sole authority of his own mind. He is no longer enchanted by the order and beauty of the whole that transcends him and places his own mind in its place.

Considering the status that metaphysics held in philosophy, religion and science for millennia the disrespect of the young generation to the old league is astonishing. Modernity was not shy in its ambitions. Descartes intended to entitle the *Discourse* "The Plan of Universal Science which is Capable of Elevating our Nature to its Highest Degree of Perfection".[7] It is true that Hume reached sceptical conclusions, finding reason a "leaky weather-beaten vessel" (T 263), but the original vision of both philosophers entailed the same basic features – to build a new system of sciences on entirely new, subjective foundations and define new rules for acquiring proper and accurate knowledge. This formidable task was considered straightforward by both Descartes and Hume. *Optimism, confidence and revolt* – these are the

[7] In his private thoughts – and after his return to metaphysics – Descartes makes the following confession: "The Lord has made three marvels: things out of nothingness; free will; and the Man who is God". Although he attributes the traditional role to God in the first two instances, he trumps them in the third.

key words of the new discourse. The new start meets the norms of a true paradigm-shift as described by Thomas Kuhn (1962) in the context of the development of science; the commitment to a certain world-view becomes shattered and the shared intellectual framework is lost; we can detect the emergence of a completely new way of thinking, with new values and methods that carry with them a new semantics.[8] From Descartes on, concepts like mind, knowledge, reason, man or nature became incommensurable with how they were understood before. This whole shift can be summarized as the turn to the Self.

1.5 The Self

Once philosophy starts from the autonomous cognitive self the problem of establishing the existence of an independent reality becomes acute. What can be knowledge about – about our mind contents or about real nature? Since mind ceased to be a part of a wider metaphysical framework it becomes impossible to justify the existence of a mind-independent reality; that would entail reaching out from the sphere of subjective mental acts to a different, material mode of existence, and to establish some channel of communication between the two. Descartes showed the futility of such speculations in the radical part of his *Meditations*, in the famous argument about a continuous dream. He doubts one alleged certainty after another and concludes that from the position of subjective consciousness all these certainties could just be dreams. After philosophical scrutiny he can no longer be certain that he is sitting by the fire, wearing a winter cloak and holding papers in his hand; he could be sleeping and having a detailed dream in which all these activities took place. When we are dreaming the contents of our dreams may seem very real and only on waking up do we realize we were fooled when we took them for real; and is this waking state not just another dream? Descartes then continues: "As though I were not a man who habitually sleeps at night and has the same impressions (or even wilder ones) in sleep as these men when they are awake" (Descartes 1979b, 62). But even if everything that I think is erroneous it is I who errs. Whatever I think, the act itself proves my existence as *res cogitans*; I must "conclude that this proposition 'I am', 'I exist', whenever I utter it or conceive it in my mind, is necessarily true" (Descartes 1979b, 67). In the *Discourse* Descartes reaches a similar conclusion yet without the precise philosophical justification, saying: "I decided to feign that everything that has entered my mind hitherto was no truer than the illusions of dreams. But immediately upon this I noticed that while I was trying to think

[8] Descartes made a breakthrough also in the area science, especially in mathematics. He applied algebraic analysis to geometry, and by enabling the visualization of algebraic problems in terms of curves he laid the foundation necessary for Leibniz and Newton's construction of calculus. For more details about Descartes' place in other revolutions in mathematics throughout history see Gillies 1995.

everything false, it must needs to be I, who was thinking … I am thinking, therefore I exist" (Descartes 1979a, 31).[9]

After rejecting all metaphysical authority and adopting the position where "I was as it were (*tanquam*) forced to become my own guide" (Descartes 1979a, 19), Descartes shifts his attention to defining the rules of scientific method. This method should ensure that knowledge achieved would have the same (or maximum possible) degree of evidence like the statement "I am thinking, therefore I exist" – the model of indubitable certainty. Descartes's normative method based on deduction and intuition was designed for natural science; it relies on the *lumen naturale* of our reason, imposing our measures upon nature. Likewise, Hume postulates methodological rules of the conduct of the mind that rests on the priority of experience, according to which all knowledge proceeds from observation; in both cases the operations of our mind that constitute knowledge. An account of the powers and the capacity of our understanding will, according to Hume, provide the foundations for all sciences – whatever the subject they depend on our reasoning. Despite these differences, Hume's starting point is the same as Descartes' – the subject of epistemology *is our consciousness and its own contents*. Hume builds on Descartes's autonomous cognitive self – *ego percipio* is an analogue of *ego cogito*.

Our subjectivity now forms a new horizon for epistemology. In Hume's words, "The only existences, of which we are certain, are perceptions, which being immediately present to us by consciousness … are the first foundation of all our conclusions" (T 212). Knowledge does not come, as before, from Being, from things that reveal themselves to us because such is the purpose of the whole but starts with the active subject and his projections of the world. Descartes cannot answer questions of whether we have the right to pass from the idea of the thing to the thing itself, or whether the clearness and the distinctness of an idea guarantees its objective *validity*, "after all, the clearness of an *idea* is one thing – and the real existence of the object of it quite another" (Koyré 1979, xxxi). Since ontology is dismissed from philosophy and epistemology takes its place the prime target of philosophy is to look at our projections that constitute our world (and its objective validity); therefore, what we *know* 'makes' what there *is* (for us), and we dignify it with the title of *realities*. It is the projective capacity of the mind that becomes the new and enduring theme of philosophy. Hume reflects this in the famous pronouncement that "the mind has a great propensity to spread itself on external objects" (T 167), a capacity that Stroud names "gilding or staining the world" (Stroud 2007).

Descartes argues in *Rule XII* that the first step in epistemology must be the knowledge of the intellect itself since *all other knowledge is dependent on it and not otherwise*. Knowledge is possible because we get to know only what we 'make' – objects of knowledge ultimately bear the seal of our subjectivity, and due to this common ground in subjectivity the possibility of knowledge is guaranteed. Hume,

[9] Hume went even further than Descartes and doubted the existence of the self. This aspect of Hume's scepticism will not be discussed here. Hume's intention was to reject the conception of mind as a substance, but he made the same mistake. He assumed that perceptions must be tied to a substratum – to a scene, a theatre in which they make their appearance.

too, acknowledges that we be certain only that perceptions exist in *our* mind. This implies a phenomenalist position by which the mind has knowledge of its own projections, though they have an objective status for us. For Hume, "nothing is really present with the mind but its perceptions or impressions and ideas, and that external objects become known to us only by those perceptions they occasion. To hate, to love, to think, to feel, to see; all this is nothing but perceive" (T 67); 'external objects' in this context refer to the fact that we naturally ascribe existence to our impressions. The domain of subjectivity cannot be transcended.

Hume echoes Descartes, for whom man is "a being that doubts, understands, asserts, denies, and is willing; … Am *I* not the very person how is now 'doubting' almost everything; who 'asserts' this one thing … and 'denies' other things … who is 'willing' to know more … 'imagines' and 'perceives' many things?" (Descartes 1979b, 70). We conclude that these acts of thought prove the existence of me thinking (or perceiving) but not the independent existence of what is being thought or perceived. It is the subject, the self, who applies his norms to 'his' objects, and takes them for independent entities. Modern philosophical discourse thus implies phenomenalism linked to the loss of the mind independent-world – this loss does not mean a denial of the existence of such world, but a denial of the possibility of determining whether it exists or not.

1.6 The Loss of the World

The autonomy of the individual mind, introduced to philosophy by Descartes, gave man unprecedented privileges. At the same time, it led to the *imprisonment of the subject in itself.* The world in its all-encompassing unity was lost and *man lost the sense of belonging to the universe*. That was the price to pay for the newly acquired freedom. "For Hume, the human being is no longer the darling, even the fallen darling, of the cosmic order, the pinnacle of a rational plan executed by a benevolent deity who built us in his own image" (Blackburn 2008, 7). Our consciousness is a boundary that cannot be transcended, and the mere existence of something external to my mind lies beyond the scope of any rationally justifiable argumentation. As described by Husserl, "empiricist skepticism brings to light what was already present in the Cartesian fundamental investigation … namely, that all knowledge of the world, the pre-scientific as well as the scientific, is an enormous enigma" (Husserl 1970, 89–90).

Hume defines this position clearly: "Let us chace our imagination to the heavens or to the utmost limits of the universe; we never really advance a step beyond ourselves, nor can we conceive any kind of existence, but those perceptions, which have appeared in that narrow compass. This is the universe of our imagination" (T 67–8). He follows up on Descartes' claim that "although [ideas] do not depend on my will, it does not necessarily follow that they proceed from external objects … perhaps there exists in me some other faculty, as yet imperfectly known to me, that generates such ideas … without the help of any external objects" (Descartes 1979b,

79–80). The same sentiment is expressed by Hume: "As to those *impressions*, which arise from the *senses*, their cause is, in my opinion, perfectly inexplicable by human reason, and 'twill always be impossible to decide with certainty, whether they arise immediately from the object, or are produc'd by the creative power of the mind, or are deriv'd from the author of our being" (T 84). On phenomenalist grounds none of these possibilities can be confirmed or rejected.

Here, Descartes and Hume parted ways. For Descartes it was imperative to establish the existence of the world and the possibility of true – not just subjectively certain – knowledge, since as a scientist he needed the world as a real object of investigations. The *Discourse* itself was planned as an extensive scientific text, with treatises on dioptric and geometry; what is today known as the 'Discourse' is in fact a brief introduction. Hume's situation was different. He was not educated in science and not particularly interested in it. Hume gave up on a proper *philosophical* solution of phenomenalism since our understanding is not equipped for such task, and turned instead to human nature, i.e. to our basic, pre-reflexive spontaneous instincts. These instincts establish the existence of objective reality and the adequacy of our knowledge beyond all doubt but without rational justification.

Descartes as a rationalist had to insist on a logically consistent solution. The only one available was to turn to metaphysics and use God as the guarantee of the existence of the world and of mathematical structure of nature that can be grasped by our reason. He thus retreated from the position of the self and ultimately rejected the revolutionary theses that have changed the direction of philosophy for centuries. This turn begins already in in the middle of the third meditation where he postulates the theory of the so-called *realitas obiectiva*, the representational reality of ideas; it is a reality of *ideas* that represents the actual reality of objects (*realitas formalis* of things). Descartes then proceeds to the ontological proof of God's existence – he does not use the phrases 'arguments for God's existence', or 'ways to God', but the term 'proof' as in demonstrative reasoning. Nevertheless, in this turn to metaphysics Descartes devalued the thesis of the autonomy of the self and claimed that we have the idea of the self from God. He goes so far as to proclaim (echoing Aristotle) that the contemplation of the Divine is the highest bliss possible; a big difference from the earlier Descartes who was so impatient to apply the new scientific method and master nature. Yet, this shift to metaphysics bore some distinctive modern features. Despite the fact that in *Rule I* he argued (in Aristotelian fashion) that the delight (*felicitas*) we find in contemplating the (divine) truth can bring us happiness, he insisted that this contemplation must not divert our attention from the search for scientific truths; philosophy must serve this end. In the fifth *Meditation* Descartes revives the old ontological concept of truth (truth is Being) and concludes: "I see plainly that the certainty and truth of all knowledge depends entirely on my awareness of the true God" (Descartes 1979b, 108). Yet again, though, he adds the phrase 'my awareness', bringing in the modern mode of thought.

Descartes' metaphysical turn has been interpreted in many ways. Did he plan to use the existence of God to justify the possibility and the truth of knowledge from the very start? He might then have used the classical form of a polemical meditation (disputation), as was common in scholastics, by which we start with acute doubts (sometimes even concerning the existence of God or the Trinity) and consider various solutions (*Ego cogito*, in Descartes' case), only to reaffirm God's existence as the source of being and truth. Or did he resort to God because no other ('better') solution was available? It seems more probable that Descartes' God is the *deus ex machina* whose purpose is to re-establish the existence of the world and its mathematical design. God is then more a divine mathematician than the traditional spiritual and moral substance, the saviour of souls; for Descartes, to sin is to err. God created a mathematically-structured world that is transparent to our mathematical reason since God could not be so malevolent as to give us the desire for knowledge without providing us with the right means of thought to grasp it. This metaphysics is very remote from the traditional scholastic kind.

Some of Descartes' contemporaries accepted the metaphysical solution as satisfactory and no longer questioned the ontological status of the world. They exploited the modern drive for knowledge and pursued science instead, leaving behind the burden of Descartes' ontological scepticism of the radical part of his *Meditations*. Leibniz assumed that mathematics makes the universe intelligible and linked together the laws of nature and the laws of reason. Though Leibniz criticized Descartes on various accounts, he adopted the rationalistic philosophy as the foundation of science. His metaphysics was also far from traditional scholastics; he invented his own conception of monads acting in pre-established harmony set by a rational divine force. Spinoza, too, elaborated his own metaphysical theory and – in opposition to Descartes' intellectualism – turned to pantheism which identifies God with nature. But nature is for him already perceived in the modern way, as a sum of matter ordered by reason. Reason then gives us access to the cosmic and moral order.

The empiricists in France and Britain did not favour metaphysics and had to confront the phenomenalist problem. Gassendi and later Huet took the middle way; they conceded that knowledge is about appearances but never doubted that perceptions come from the outside, from objects. Gassendi, a renowned scientist, tried to link our knowledge to real objects, by placing weight on modesty and sticking to probability. Locke, Berkeley and Hume were less involved in science; they concentrated on the genetic aspects of knowledge – how our mind works: how our ideas are produced and how our beliefs arise. Locke naïvely presumed that primary qualities come from external things as imprints in our passive and empty mind. Berkeley perfectly understood the phenomenalist implications of the position of the autonomous self and denied the possibility of reaching anywhere beyond the mind (*esse est percipi*); for the proof of the existence of any corporeal or material substance, "it would be necessary that you conceive them existing unconceived or unthought of, which is a manifest repugnancy. When we do our utmost to conceive the existence of external bodies, we are all the while only contemplating our own ideas" (Berkeley 2004, 61).

Hume's place within early modern philosophy is unique. *He was the only philosopher who took up the challenge of radical Cartesian scepticism and the resulting phenomenalism with full awareness of its consequences and without any recourse to metaphysics*. Hume acknowledged that there were no arguments, either positive or negative, giving an answer to the question of the existence of anything independent of the mind; therefore, we have to remain sceptical and suspend our judgment on this issue. Phenomenalism, to define it precisely, raises the problem of the justification of our beliefs about physical world; "objects" are reduced to appearances (*phainomena*), to objects of our awareness. Anything beyond the sphere of subjectivity is not philosophically accessible. External objects as the cause of our perceptions are reduced to "a certain unknown, inexplicable *something*; a notion that is so imperfect, that no sceptic will think it worthwhile to contend against it" (E 155). In Hume's philosophy, *phenomenalism defines the deepest level of scepticism that is independent of his empiricism* and reflects the mystery concerning the objective existence of the world, the mystery that came into existence with Descartes. Husserl precisely describes this position: "Through Berkeley's and Hume's revival and radicalization of the Cartesian fundamental problem, dogmatic objectivism was … *shaken* to the foundations. This is true not only of the *mathematizing objectivism,* so inspiring to people of the time, which actually ascribed to the world itself a mathematical-rational in-itself … it was also true of the *general objectivism* which had been dominant for millennia" (Husserl 1970, 90).

In the *Treatise* Hume claims that "the first kind [of impressions] arises in the soul from unknown causes" (T 7); he confirms this claim in the *Enquiry*: "by what argument can be proved, that the perceptions of the mind must be caused by external objects, entirely different from them, though resembling them (if that be possible) and could not arise either from the energy of the mind itself, or from the suggestion of some invisible and unknown spirit, or from some other cause still more unknown to us?" (E 153). Here, Hume echoes Locke conviction that our knowledge is about appearances, "it [the object] is merely a supposed I know not what" (Locke 2011, 305). Hume already built on Descartes' claims as formulated in his *Meditations* where Descartes admits there is no solution to scepticism concerning ontological realism; he corrects the evidence that he felt earlier about various things: "Earth, stars, and the rest of what I got from the senses. Now what did I clearly perceive about them? Only that the ideas or thoughts (*cogitationes*) of such things occurred in my mind" (Descartes 1979b, 77). But Hume rejected Descartes' recourse to the veracity to God in order to establish the connection between mind and external reality; against Locke he emphasizes that all sensible qualities – not only secondary but primary as well – are in the mind, not in the object; and he rejected Berkeley's extraordinary combination of phenomenalism and religious metaphysics in which Berkeley turned to God as the source of our ideas and the guarantee of the possibility of knowledge. It was only Hume who took phenomenalism – that is scepticism with regard to the existence of the external world – to its utmost limits.

Hume, of course, proposed a solution to this scepticism. Nature saves us by implanting in us an *irresistible instinct* to believe in the existence of bodies and in the necessary connection between cause and effect. Belief in necessary connection encourages inductive expectations, i.e. our belief that the future of which we have no experience will resemble the past. Although inductive expectations cannot be *reduced* to necessary connections, the experience that fire has always produced heat makes us believe that fire *must* always produce heat. Yet this extrapolation from the observed to the unobserved is based on instincts, not on rational reflection.

Hume was the only philosopher of his times who expunged all metaphysics from epistemology and had the courage to hold a consistently phenomenalist position. The question "what is beyond our mind?" becomes meaningless for Hume – some unknown something is not worthy of our attention. Our beliefs and imagination are of course unaffected by this scepticism but are beyond the jurisdiction of reason. Using the old metaphor of light almost in a Platonic sense, Hume says that "she [reason] sees full light, which illuminates certain places; but that light borders upon the most profound darkness. And between these she is so dazzled and confounded, that she scarcely can pronounce with certainty and assurance concerning any one object" (E 157).

1.7 The "New Hume"

From the perspective of Hume's phenomenalism it seems interesting to consider the position labelled the 'New Hume' (NH) interpretation. It addresses the issues of projectivism and realism in Hume's epistemology and opens a discussion about Hume's possible ontological commitments. The NH debates have grown extensively in the last decades producing a great diversity in opinions. I have to restrict my analysis to one key claim introduced by the 'founding fathers' of NH, John P. Wright (1983) and Galen Strawson (1989). They disagree with the 'standard' (Old) Hume position that apparently claims that as a phenomenalist and projectivist Hume denied the real existence of objects and causation. In Strawson's words, the epistemological claim "All we can ever know of causation is regular succession" is *catastrophically* (*sic*) extended to the ontological claim, "All that causation actually is, in the objects, is regular succession", a move he finds fantastically implausible (Strawson 1989, 11; 2007, 33). According to New Humeans, the standard interpretation of Hume's scepticism entails the ontological claim that causation is nothing but a regularity theory and that objects are nothing but perceptions.

If this criticism of the standard interpretation is correct the 'Old Humeans' must be accused of several things: of bringing ontology (in the negative sense) to Hume's philosophy though Hume himself discarded this whole area of inquiry as meaningless; of making Hume a dogmatic since a modern sceptic could not make claims about the existence (or non-existence) of anything beyond our experience; and of presenting Hume as an idealist and an extreme subjectivist. Before making such verdict several issues should be addressed: firstly, is it not the NH interpretation that makes ontological claims – asserting that objects and causation really exist? In that case, the New Humeans could be blamed for the same 'vices' for which they blame the traditional Humeans, especially for drawing Hume into metaphysics and dogmatism. Secondly, if the New Humeans do not make any ontological claims what innovation do they actually bring into Hume scholarship? And thirdly, does the standard interpretation (*en bloc*) really assert that Hume denied the extra-mental existence of objects or causal relation?

I shall start with the first question. Given the fact that the NH philosophers call Hume's position 'sceptical realism' (as in the title of Wright's book) the ontological thesis offers itself; while Hume was an epistemological sceptic he was a *realist* with regard to causation and objects. Wright says that for Hume "the subjective character of our perceptions presupposes the existence of independent external objects which are related to those perceptions" (Wright 1983, 40), and makes a similar statement with regard to causation: "[Hume] constantly maintained that here are real powers and forces in nature which are not directly accessible to our senses" (Wright 1983, 129). To soften the impact of this claim he then argues that our ideas are only inadequate representations of reality (since they cannot be broken down to corresponding impressions). Strawson holds a similar view; "objects", he says, "are genuinely non-mental, things that exist independently of our minds"; similarly, Causation (with a capital "C") is "something essentially more than regular succession", and although the ultimate springs of causation are totally shut up from human curiosity, "they certainly exist" (Strawson 2007, 33). But of course we cannot understand the true nature of these powers. This sceptical realist picture of Hume is defined by Kenneth Richman in the 'Introduction' to the *New Hume Debate*, a volume that discusses the views of the most prominent NH supporters and opponents. He says: "A sceptical realist about some entity is realist about the entity's existence, but agnostic about the nature or character of that thing because it is epistemically inaccessible to us" (Read and Richman 2007, 1). The argument of the NH proponents is based on a list of Hume's references to hidden powers, energy, force or efficacy that cannot be comprehended – they are concealed from us and we cannot get acquainted with them – nevertheless they have real existence; to emphasize the ontological status Strawson uses capital letters for Existence and Causation.

This indicates that the *NH interpretation does make positive ontological claims*. It is also true that many of Hume's own statements do encourage this interpretation; most famously, to give at least a couple of examples, the one about the secret powers that nature conceals from us but on which the influence of objects entirely depends (see E 33), or about "the power or force which actuates the whole machine [the universe] yet it is entirely concealed from us, and never discovers itself in any of the sensible qualities of body" (E 63). But by the same token, Hume often argues the opposite: the efficacy or energy of causes belongs entirely to the soul, it lies in the determination of the mind, arises from reflecting on the operations of our own mind, and is tied to the propensity of mind to spread itself on external objects; the term 'external objects' here means, to repeat, objects constituted by our mind that we consider external (see T 166–7, E 33, 63–4). Obviously, it is the task of a contextual interpretation to provide a unified picture of Hume's views on this matter.[10]

[10] Peter Millican provides an exhaustive analysis of the NH realist claims and argues that the NH quotes are not only outnumbered by those that do not allow the realist interpretation, but are often misconstrued. More importantly he demonstrates that the realist picture of Hume does not fit into

This is not the place to pursue the NH debate in detail. The task here is to offer another perspective for assessing Hume's position which takes into account the Cartesian roots of his scepticism. In this perspective the NH ontological claims about Hume's philosophy seem misleading in principle. As explained in the previous sections, ontological inquiries are explicitly made redundant by Hume since *nothing can be asserted about any reality external to the mind*; "nothing can be ever present to the mind but the image or perception, and senses are only inlets, through which these images are conveyed, without being able to produce any immediate intercourse between the mind and the object" (E 152). Hume, as a philosopher of the modern era, and the one who abandoned all ties to metaphysic, invested all his energy towards developing a theory of knowledge based on the observation of the mechanisms of the individual mind – that was the new sexy area of inquiry. I agree with Winkler's view that "the issue of causal [or any] realism was unimportant for Hume – a peripheral issue he had no real need to clarify" (Winkler 2007, 72).

Hume abandoned the pursuit of the 'WHAT' (ontological) questions which, as he repeatedly declared, lead to absurdities and to metaphysical speculations. They are not only improper but also unimportant and even harmful – "obscurity and error begin to take place, and we are led astray by a false philosophy" (T 168). Hume's advice is clear: "if it [the philosophical enterprise] lies within the compass of human understanding, it may at last be happily achieved; if not, it may, however, be rejected with some confidence and security" (E 15). Hume had no regrets about the loss of ontology since he replaced it by a more intriguing epistemological inquiry about HOW our mind works; the first task is to get "thoroughly acquainted with the extent and force of human understanding … and with the operations we perform in our reasoning" (T xv). Hume was interested in the cognitive process not in its ontological foundations.

What is, then, the benefit of searching for some signs of realism in Hume's epistemology? This line of interpretation seems uninspiring, focusing on issues that were *passé* for Hume, belonging to the old metaphysical tradition that was now, as he claimed, superseded by the investigation of human nature. The abstruse philosophy – of which the ontological questions are part – represents the airy science of the old wisdom whose foundations are rotten. Hume shifts the epistemological discourse to studying the cognitive nature of the mind, including our imagination and creativity. The hunt for realism diverts attention away from the most original content of Hume's philosophy: the most radical elimination of metaphysics and most consistent phenomenalism among his peers, opening new domains of philosophical inquiry.

Searching for ontological realism in Hume's philosophy seems misguided in the same way as attributing ontological realism to Kant would be – even though Kant begins his first *Critique* with a clear realist statement; "There can be no doubt that all our cognition begins with experience. For what else might rouse our cognitive power to its operation if objects stirring our senses did not do so?" (Kant 1996, B 1,

---

the philosophical context of Hume's writings from the *Treatise* and the *Abstract* to the *Enquiry* (Millican 2007a, b).

43). Hume did not go that far and remained noncommittal; the 'stirring' may be due to some power inside me, yet unknown to me. But would any scholar call Kant a realist? Clearly not, since realism does not epitomize Kant's contribution to epistemology – his achievement lies elsewhere, in the conception of transcendental subjectivity as the foundation of objectivity (for us). To quote Husserl again, "it is a philosophy which, in opposition to prescientific and scientific objectivism, goes back to knowing subjectivity as the primal locus of all objective formations of sense and ontic validities" (Husserl 1970, 99). Not realism but by contrast a philosophy that is concerned "not so much with objects as rather with our way of cognizing objects in general" is the center of investigations; cognizing means, of course, constituting since we can get knowledge only about something that bears our imprint (Kant 1996, B 25, 64).

But is the ontological claim of the NH position as outlined above not unfair? This takes to the second question. Looking at Strawson's analysis one can find weaker claims, defining the NH position in a more minimalist way. Strawson argues that New Humeans only reject the "Old", allegedly ontological interpretation according to which Hume *denies* the existence of anything external to the mind. According to Strawson, Hume holds "strictly non-committal scepticism with regard to knowledge claims about the nature of reality – his strictly non-committal attitude about questions about what we can know to exist *or know not to exist*, in reality" (Strawson 1989, 11). In case of causation, we cannot *rule out* that there is such a thing as natural necessity or causal power; "at no point in the *Enquiry* … does he [Hume] even hint at the theses that all there is to causation in the world is regular succession" (Strawson 2007, 48). In case of objects, Hume "certainly does not mean to imply that there are no external objects" (Strawson 1989, 52). This account fits perfectly with Hume's scepticism entailing that we cannot prove or disprove anything beyond our experience, beyond appearances. However, this account is not consistent with Strawson's ontological claims on Hume's behalf described above, and his position is confusing as a result.

If we were to accept the softer NH version of Hume's realism the question arises whether it is realism (Realism) at all, and whether it is legitimate to use the terms Causation and Existence in the ontological sense. Within the diversity of the NH interpretations there are many attempts to avoid the contradiction between the realist ontological commitment and projectivism.[11] For instance, Hume's alleged realism is defined in terms of mere realist *assumptions* or *suppositions*; we cannot conceive of necessary connection but we can suppose it exists in nature. Supposing must not be confused with spontaneous belief or imagination but it is, as Strawson

[11] For instance Edward Craig (2007) argues that Hume can be interpreted in both the realist and the anti-realist way and that there does not have to a collision between the two views. Peter Kail (2007), too, is cautious about Hume's realism and admits that Hume's stance is not univocal; he admits that two strains of thought – the realist and the projectivist – can be found in Hume's epistemology and attempts to reconcile them. Helen Beebee (2006, 2007) identifies three interpretative positions relevant for Hume's conception of causation, the traditional, the sceptical realist and the projectivist (which is closer to the traditional view) and argues that Hume's writings allow for all these alternative readings.

argues, a philosophical attitude. The supposition of the existence of some mind-independent reality is seen as a solution to the problem of our having ideas without having corresponding impressions; this solution enables us to form 'relative ideas' (without pretending to comprehend the related objects). But, as Winkler points out, at Hume's time the use of the word "suppose" was very liberal; we can suppose anything, even contradictions and absurdities (Winkler 2007, 64). In any case, according to this mitigated version of realism "Hume will not claim that we can know that there is definitely nothing like Causation in reality. Equally, though, he will not claim that there definitely is something like Causation in reality" (Strawson 2007, 34). This conclusion is perfectly consistent with Hume's *modern* scepticism. A modern sceptic must suspend judgment (not belief), positive or negative, on the mere existence of any reality beyond appearances (unlike a Pyrrhonian who doubts only the possibility to get the knowledge of the intrinsic nature of things behind appearances). Modern scepticism thus entails phenomenalism as its integral part; but phenomenalism in no way denies the existence of some extra-mental physical reality and laws of nature. But this conclusion is also consistent with most standard Hume interpretations.

Their alleged novelty of the NH is based on the assumption that the standard (Old) Hume interpretation makes an ontological claim (to the negative effect), namely that it holds a pure regularity theory of causation and a purely subjectivist theory of objects (defined as mental entities); therefore, the NH say, the standard interpretation of Hume denies the real existence of objects and causality in nature. But, to raise the third question, is this evaluation of the previous tradition of Hume scholarship correct? Unfortunately, the claims of the NH position are not backed by sufficient evidence. Strawson neither analyzes the relevant traditional texts nor specifies who exactly should be included in the standard group; instead, he uses phrases like "it has been widely believed", or "some doubt", or "some tend to think".[12] But if we take a look at the views of two prominent Hume scholars, Anthony Flew and John Passmore we find that they do not make any ontological claims about Hume. Flew says that Hume "cannot know there are mind-independent objects in an External World. … The true and inescapable outcome is epistemological solipsism; no one can know that the Universe contains anything or anyone other than himself (or herself, or itself)", and continues on the next page: "we are never so privileged as to be granted any such close contacts with the External World" (Flew 1986, 30–31). Not only is there no 'catastrophic' transfer of epistemological scepticism to ontology but Flew – referring to the influence of Descartes on Hume – says: "we are each and all of us separately and individually shut off from any External World which may or may not exist", and therefore, he continues, "he [Hume] is entitled to employ the word 'experience' … with reference only to ongoings in his own mind – to his Internal World, so to speak" (Flew 1986, 15, 31).

[12] In the Preface to his book (1989) Strawson refers to Anthony O'Hear (1985) and to Robert Schacht (1984), but the first book is an introduction to Hume's philosophy not intended for Hume's scholars, and the second is a textbook of moderns philosophy.

We find no denial of the possibility of the existence of any external world but are advised by Hume (Flew says) not to pursue any investigations about this matter.

Passmore discusses various of Hume's intentions and in the section on Hume's phenomenalism (the theme most exposed to the NH criticism) draws attention to Hume's statement that we cannot *know* anything but perceptions; therefore, we cannot *prove* that bodies exist; this question is unanswerable and "we should, for this question, substitute the *psychological* problem: 'How do we come to believe that bodies exist?'" (Passmore 1952, 91). This approach is similar to Flew, and a reasonable one. No assertive verdict about Externality is allowed in the region of philosophical reflection. The fact that the immediate objects of thought as well as the relations between or among those objects occur in our mental world does not imply that there is or is not an extra-mental world that causes our perceptions. As Norton remarks, "[Hume's] concern is not to advance from this base in order to deny dogmatically that there are causes, objects, or minds" (Norton 2009, 10–12). The standard interpretation thus seems compatible with the weaker claims of the NH proponents; in that case, though, these New Humeans should not consider themselves realists. Moreover, they would not claim anything in principle different from the standard interpretation.

Some examples of a pure projectivist ('anti-realist') reading of Hume can be also found in the traditional interpretation. To name but one, Stroud (1977, 2007) argues that objects do not really stand in causal relations but our imagination adds the 'new item' in the mind. But even such statements can offer a non-committal reading if considered in the Cartesian context. In this context our subjectivity is the limiting horizon of our inquiries; if we do not rely on metaphysics the question of what there is becomes identical with the question of what we know. Projectivist readings may thus dissolve ontology in epistemology but they do not transfer epistemology to ontology, as Strawson assumes. Statements like "there is not any more to causality than regularity of succession (or constant conjunction)" do not necessarily have ontological relevance but express the fact that we only have access to the world as we know it; this world is relevant for our lives and cognition, and deserves philosophical attention. So, instead of accusing the standard interpretation of seeing Hume as denying the existence of real powers the New Humeans should consider the opposite – namely "that Hume's scepticism may consist in a *refusal to affirm* the existence of real powers", as Winkles observes (2007, 67). For Hume, we cannot reach the other – unobservable – side of our perceptions. The other side is "mysterious and straightjacketing" and philosophical understanding "must sail on in complete indifference to any facts that transcend our ideas", as Blackburn puts it (2007, 106). For Hume, ontological inquiries lead to metaphysics, and "metaphysics arise[s] … from the fruitless efforts of human vanity, which would penetrate into subjects utterly inaccessible to the understanding" (E 11). The NH interpretation seems to steer the discussion of Hume's epistemology exactly to this domain.

How, then, can one resolve the tension in Hume's epistemology between the rule that we cannot form an idea without a corresponding impression, and the fact that we do have such ideas and that they are intelligible? Hume is not helpful on this task. On the one hand he states that "it is not possible for us ever to form the most distant idea of it [necessity]" (T 165). On the other hand we have this idea and use it in all our inferences. The answer could be that *we do not have ideas of causality or a body*; not the *proper* or *adequate* ideas formed as prescribed by the copy principle. We naturally (and ignorantly) "confound ideas, which are entirely distinct from each other" (T 168) and imagine (and believe) more that our experience allows. NH supporters put a strong emphasis on Hume's sentence that there may be some qualities in material objects – but the sentence rather shows the irrelevance of such concerns for Hume: "if we please to call these *power* or *efficacy*,'twill be of little consequence to the world" (T 168). After failing in providing a philosophical solution Hume ultimately turns to our natural inclinations. We simply have the thoughts and use the words because that is how we talk and what we believe. If we discard sceptical reflection and let us be carried along by natural instinct we get more than just inadequate representations of reality (or relative ideas): we get full blown reality free of scepticism but also free of philosophy. We employ these words in everyday sense, using ordinary language as the only way to express ourselves. Berkeley could provide some inspiration here (Berkeley 2004, 71):

> In the ordinary affairs of life, any phrases may be retained, so long as they excite in us proper sentiments, or dispositions to act in such a manner as is necessary for our well-being, how false soever they may be if taken in a strict and speculative sense. Nay, this is unavoidable, since, propriety being regulated by custom, language is suited to the received opinions, which are not always the truest. Hence it is impossible, even in the most rigid, philosophic reasonings, so far to alter the bent and genius of the tongue we speak, as never to give a handle for cavillers to pretend difficulties and inconsistencies. But, a fair and ingenuous reader will collect the sense from the scope and tenor and connexion of a discourse, making allowances for those inaccurate modes of speech which use has made inevitable.

References

Beebee, H. 2006. *Hume on causation*. Abingdon: Routledge.

Beebee, H. 2007. The two definitions and the doctrine of necessity. In *Proceedings of the Aristotelian Society*, Vol. CVII, 413–431.

Berkeley, G. 2004 [1710]. *A treatise concerning the principles of human knowledge*, ed. R. Woolhouse. London: Penguin Books.

Blackburn, S. 2007. *Hume and thick connections*. In ed. R. Read and K.A. Richman, 100–113.

Blackburn, S. 2008. *Hume*. London: Granta Publications.

Craig, E. 2007. Hume on causality: Projectivist and realist? In ed. R. Read and K.A. Richman, 113–122.

Descartes, R. 1979a [1637]. Discourse on method. In *Descartes. philosophical writings*. Trans. E. Anscombe and P.T. Geach, 5–59. Nelson's University Paperbacks.

Descartes, R. 1979b [1641]. Meditations on first philosophy. In *Descartes' philosophical writings*. Trans. E. Anscombe and P.T. Geach, 59–125. Nelson's University Paperbacks.

Descartes, R. 1985 [c. 1628]. Rules for the direction of the mind. In *The philosophical writings of Descartes*, vol. 1. Trans. J. Cottingham, R. Stoothoff and D. Murdoch, 9–79. Cambridge: Cambridge University Press.

Flew, A. 1986. *David Hume. Philosopher of moral science*. London/New York: Basil Blackwell.

Gillies, D. (ed.). 1995. *Revolutions in mathematics*. Oxford: Clarendon.

Hegel, G.W.F. 1894 [1837]. *Lectures on the history of philosophy*, vol. II. Trans. E.S. Haldane and F.H. Simson. London: Kegan Paul, Trench, Trübner and Co.

Hesiod. 1914 [c. 6 BC]. *Theogony*. Trans. Evelyn-White, H.G. Harvard University Press.
Husserl, E. 1970. *The crisis of European science and transcendental phenomenology*. Trans. D. Carr. Evanston: Northwestern University Press.
Kail, P.J.E. 2007. *Projection and realism in Hume's philosophy*. Oxford: Oxford University Press.
Kant, I. 1996 [1781]. *The critique of pure reason*. Trans. W.S. Pluhar. Indianapolis: Hackett Publishing Company.
Koyré, A. 1979. Introduction. In *Descartes. philosophical writings*. Trans. E. Anscombe and P.T. Geach, vii–xlv. Nelson's University Paperbacks.
Kuhn, T. 1962. *The structure of scientific revolutions*. Chicago: University of Chicago Press.
Locke, J. 2011 [1689]. An essay concerning human understanding, ed. P.H. Nidditch. Oxford: Oxford University Press.
Millican, P. 2007a. Against the 'New Hume'. In ed. R. Read and K.A. Richman, 211–253.
Millican, P. 2007b. Humes old and new: Four fashionable falsehoods, and one unfashionable truth. In *Proceedings of the Aristotelian Society*. Supplementary vol. LXXXI, 163–199.
Norton, D.F. 2009. An introduction to Hume's thought. In *The Cambridge companion to Hume*, ed. D.F. Norton and J. Taylor, 1–40. Cambridge: Cambridge University Press.
O'Hear, A. 1985. *What philosophy is*. Harmondsworth: Penguin.
Passmore, J. 1952. *Hume's intentions*. London: Cambridge University Press.
Read. R., and K.A. Richman, ed. 2007. *The New Hume debate*, rev. ed. London/New York: Routledge.
Schacht, R. 1984. *Classical modern philosophers. Descartes to Kant*. London/New York: Routledge.
Strawson, G. 1989. *The secret connexion: Causation, realism, and David Hume*. Oxford: Oxford University Press.
Strawson, G. 2007. David Hume: Objects and power. In ed. R. Read and K.A. Richman, 31–52.
Stroud, B. 1977. *Hume*. London: Routledge & Kegan Paul.
Stroud, B. 2007. 'Gilding or staining' the world with 'sentiments' and 'phantasms'. In ed. R. Read and K.A. Richman, 31–52.
Winkler, K.P. 2007. The new Hume. In ed. R. Read and K.A. Richman, 52–87.
Wright, J.P. 1983. *The sceptical realism of David Hume*. Manchester: Manchester University Press.

Chapter 2
The Empiricist Roots of Hume's Scepticism

Abstract Hume abandoned all questions concerning the origins of perceptions and shifted his attention to internal mental processes. Epistemology is concerned with mapping the mind and Hume thus delineated a new area of philosophical inquiry that later led to the emergence of disciplines like the philosophy of mind and cognitive science. Yet these inquiries brought to light another level of scepticism. Empiricism proceeds from the rule that all knowledge must be based on observation and experience. But neither the idea of causation nor distinct existence can be derived from experience; they are based on instincts and imagination, not on reason. This conclusion seems damaging for the status of science. At that time of advancing modernity, culminating in the Enlightenment confidence in reason and cognitive optimism, such a conclusion was very disturbing indeed. Epistemology was supposed to provide rational foundations for science and this was what Hume could not deliver. Hume, however, was not concerned about natural science. He appreciated Newton's *method* and presumed that by its application to human sciences this new experimental approach would be enriched, bringing new and much needed discoveries about *human nature*.

Keywords Empiricism • Causation • Science • Reason • Instinct

2.1 From Perceptions to Ideas

Hume's epistemology is based on empiricism and on its basic principle that *all knowledge is derived from experience and must not go beyond experience*. Hume formulates this principle both in the *Treatise* and the *Enquiry*; for example, he says, "and as the science of man is the only solid foundation for the other sciences, so the only solid foundation we can give to this science itself must be laid on experience and observation" (T xvi), and he continues, "'tis still certain we cannot go beyond experience; and any hypothesis, that pretends to discover the ultimate original qualities of human nature, ought at first to be rejected as presumptuous and chimerical" (T xvii). Hume joined the group of empiricists that initiated early modern philosophical and scientific discourse, represented most notably by the French empiricists Pierre Gassendi and Pierre Daniel Huet; and in Britain by John Locke and

Z. Parusniková, *David Hume, Sceptic*, SpringerBriefs in Philosophy,
DOI 10.1007/978-3-319-43794-1_2

George Berkeley, plus "the greatest and rarest genius that ever arose for the ornament and instruction of the species", Isaac Newton (Hume 2005, 374). The empiricists stood, broadly speaking, against the rationalists (e.g. Descartes and Leibniz), who mostly employed metaphysical assumptions and considered experience an unreliable factor in the acquisition of knowledge. Despite many individual differences, empiricism was the trademark of the new experimentally-based science and a new no-nonsense philosophy rid of obscurity, speculation and superstition.

Hume's conception of knowledge acquisition is broadly concerned with the mechanisms of deriving ideas from perceptions, and tying them together; in this respect Hume was inspired by Locke who claims: "All Ideas come from Sensation or Reflection. Let us then suppose the Mind to be, as we say, white Paper, void of all Characters, without any Ideas:—How comes it to be furnished? … To this I answer, in one word, from Experience. In that all our Knowledge is founded; and from that it ultimately derives itself" (Locke 2011, 62). Hume's specific description of these operations differs from Locke's but both stand in opposition to Descartes' theory of innate ideas and rational intuition; they reject the argument that we can establish certainty regarding knowledge through the use of pure reason (or 'the light of reason'), enabling us to measure the degree of certainty against the absolute certainty of *Ego cogito*. As Hume comments on Locke, "the first proposition he advances, is, that all our ideas, or weak perceptions, are derived from our impressions, or strong perceptions, and that we can never think of any thing which we have not seen without us, or felt in our own minds. This proposition seems to be equivalent to that which Mr. *Locke* has taken such pains to establish, *viz. that no ideas are* innate" (T *Abstract* 647).

Perceptions are the basic building blocks of the mental world and also the total contents of our minds; in Hume's epistemology they play the same role as atoms in Newton's (and earlier in Gassendi's) science of nature. To précis Hume,[1] perceptions can be divided into two classes – impressions and ideas. Impressions include sensing (hearing, seeing) and feeling (pain, pleasure) and differ from ideas in their greater force and vivacity. Ideas account for our ability to think and are faint images of impressions: "This distinction", Hume says, "is evident; as evident as that betwixt feeling and thinking" (T *Abstract* 647). However, these distinctions may not be as evident as Hume claims; he often uses vague criteria, for instance when he talks about simple and complex impressions and ideas, or about belief – we know what it is when we experience it, and no definitions formulated aloof from this world can capture our psychological conviction in an adequate way. Hume then proceeds to make more subtle distinctions between impressions. There are impressions of sensations, derived from the senses; though we do not know their original cause they provide the basic material of our mental life. Impressions of reflection are produced in the reverse order from ideas, and usually concern our bodily feelings or strong emotions (e.g. pain, pleasure, hope, fear or thirst); for instance the experience of

[1] For a more detailed analysis of Hume's account on the operations of the mind see, for instance, Simon Blackburn (2008), David Owen (2009) and Don Garrett (2011).

nausea caused by eating a rotting fish is stored as an idea in the memory – and if we recall it, perhaps in seeing a fish, it is as if we feel the nausea again.

Further, Hume distinguishes between simple and complex impressions and ideas. Simple ideas are copied from single impressions and complex ideas are combinations of simple ideas. Complex ideas are therefore not copied from complex impressions which may thus seem somewhat redundant; yet they have a role to play in the composition of our nature. To illustrate by way of examples, simple impressions are perceptions of single colours, tastes or shapes; whereas complex impressions unite more *specific* simple impressions – say, the colour, shape, smell, texture and taste of an apple; they are compounded from the simplest sensory data. Simple ideas are derived from simple impressions; returning to the apple, simple ideas would be ideas of roundness, redness (or yellowness, or greenness), sweetness (or sourness). Complex ideas are compounds of simple ideas, for instance an idea of an apple on a more general level; it does not have to contain the exact colour or taste of a particular apple but retains more general characteristics such as being round, hard, red (or green or yellow), etc. We can create complex ideas of non-existent fantasy creatures such as a dragon or a unicorn from simple ideas by binding them together arbitrarily in new, unnatural combinations. Overall, Hume constructs a scheme of mental operations in which knowledge proceeds from sense perceptions and impressions to ideas. This empiricist account of knowledge is encapsulated in the *copy principle*: "*all our simple ideas in their first appearance are deriv'd from simple impressions, which are correspondent to them, and which they exactly represent*" (T 4). This principle also determines the space in which our thought is supposed to operate.

How does it come that simple ideas compound together to form complex ideas? Hume ascribes this function to *associations*: "were [simple] ideas entirely loose and unconnected, chance alone wou'd join them; and 'tis impossible the same simple ideas should fall regularly into complex ones (as they commonly do) without some bond of union among them, some associating quality, by which one idea naturally introduces another" (T 10). Hume names three kinds of associations – resemblance, contiguity and cause and effect – which are the uniting principles among ideas; as Hume says, they are "the only ties of our thoughts, they are really *to us* the cement of the universe, and all the operations of the mind must, in a great measure, depend on them" (T *Abstract* 662). Upon closer examination we shall find, Hume says, "that all this creative power of the mind amounts to no more than the faculty of compounding, transposing, augmenting or diminishing the materials afforded us by the senses and experience" (E 19). His epistemology is thus concerned with our reasoning concerning matters of fact which may, in principle, be experienced; matters of fact are objects in the empirical world or, rather, in what we naturally take to be the real world. Hume contrasts reasoning about matters of fact with relations of ideas which are the objects of mere operations of thought; these are mathematical entities that "are not existent in the universe" (E25) and of which we are demonstratively certain.

Hume considers the effect of associations equivalent to the force of attraction (i.e. gravity) in Newton's physics; both are the 'cement' of the universe, of the physical and mental realms respectively. Due to associations, ideas form an intelligible network of thought – and are the fabric of *our* intelligible world. In the *Abstract*, published anonymously as a review of his own *Treatise*, Hume considered the theory of associations his greatest contribution to philosophy, for which he deserves "so glorious a name as that of an inventor", and thus placed himself alongside Newton. (*Abstract* T 661). Nevertheless we can ask whether this description of mental life may be too narrow; Hume drew the map of mental life – based, as he argued, on careful observation of its particular components – at the age of 27 and never made any attempts to broaden it; he even simplified it in the *Enquiry*. He considered the three kinds of associations 'compleat', perhaps to make the number correspond to Newton's three laws of motion, these associations alone being responsible for forming ideas and connecting them in a train of thought.

Hume was confident that he had managed to establish the foundations of a fresh, science-friendly philosophy devoid of speculation and flights of fancy. There is just one simple recipe – "we need but enquire, from which impression is that supposed idea derived … by bringing ideas into so clear a light we may reasonably hope to remove all dispute, which may arise, concerning their nature and reality" (E 22). However, Hume's own further investigations proved that far from removing the propensity for disputes, debate increased, leading to scepticism about the rational legitimation of knowledge, thus undermining the foundational role of the copy principle. Upon closer examination we find that our natural cognitive faculties are not restricted to experience but make universal claims that cannot be derived from experience; we instinctively make a 'leap' of faith from observed instances (in the past and present) to the unknown future and believe that the sun will rise tomorrow, despite this expectation lacking any rational basis. This scepticism arises most sharply in rational reflection on the ideas of necessary connection and distinct existence since these 'ideas' are formed by our imagination but lack corresponding impressions. Imagination produces beliefs that are more vivid and forceful than mere imaginings but are "not founded on reasoning, or any process of the understanding" (E 32), and are therefore *fictions*. Fictions of this kind do not describe some 'fantasy' creations like a unicorn; after all, the idea of a unicorn may be reduced to various simple ideas and simple impressions. However, fictions of continued existence, causal connection or belief that the future will resemble the past cannot be so disassembled because there is no original impression from which they were derived. This conclusion takes Hume straight into sceptical contradictions.

2.2 Hume and Newton's Experimental Method

Hume, like most of his contemporaries, had a great admiration for Newton; as I said, Hume was inspired by Newton's experimental method. This inspiration is reflected in the subtitle of the *Treatise*, formulated by Hume as an *Attempt to Introduce the Experimental Method of Reasoning into Moral Subjects*; the determination to construct philosophy in the spirit of science was not unusual at that time.

One part of the subtitle to Hutcheson's *Inquiry Into the Original of Our Ideas of Beauty and Virtue* defines it as an *Attempt to introduce a mathematical calculation in subjects of morality*; Spinoza referred to his *Ethics* as "*ordine geometrico demonstrata*". Even philosophers who were not directly concerned with science (i.e. natural philosophy) liked to use the method and forms of science, showing their allegiance to the new age and scientifically-oriented discourse based on rational argument, empirical evidence and freedom from scholastic dogma. Newton was, of course, the most celebrated figure of the age, a discoverer of the laws and forces which govern nature and the universe. Newton's universal theory of gravitation and his laws of motion, mathematically formulated, were taken to be fundamental truths. Scientists believed that at long last, after vast tracts of time spent in ignorance and guesswork, finally the workings of the earthly and celestial worlds had been revealed.

The Royal Society, officially founded in 1663 with the assistance of Robert Boyle and John Wilkins, became the centre of experimental science: its reputation peaked under Newton's presidency, from 1703 until his death in 1727. The motto of the Royal Society, *Nullius in verba*, which means *take no one's word for it*, expressed a determination to reject any and all dogmas and superstitions and to accept only knowledge that had been experimentally tested and confirmed by experience. The intellectual world was simply fascinated by Newton. Locke, from 1668 also a Fellow of the Royal Society, claimed that in an age which had produced such Masters as the incomparable Mr. Newton, "it is Ambition enough to be employed as an Under-Labourer in clearing the Ground a little, and removing some of the Rubbish, that lies in the way to knowledge" (Locke 2011, 10). Alexander Pope glorified Newton in his epitaph (which, however, was not placed on Newton's monument at Westminster Abbey); the most famous verse says: "Nature and Nature's laws lay hid in night; God said, Let Newton be! and all was light".

Hume did not use such ostentatious phrases but he repeatedly expressed his indebtedness to Newton; he considered himself to be the 'Newton of human sciences', with the aim of bringing a similar contribution to mankind in his science of man. Newton, as Hume says, "determined the laws and forces, by which the revolutions of the planets are governed and directed. The like has been performed with regard to other parts of nature. And there is no reason to despair of equal success in our enquiries concerning the mental powers and economy, if prosecuted with equal capacity and caution" (E 14). Hume was determined to apply Newton's method, the method which forbids any speculations, sticks to what is observable and gives up the search for any secret, ultimate sources of phenomena. This attitude is often linked to one passage of Newton's *General Scholium*: "hitherto I have not been able to discover the cause of those properties of gravity from phænomena, and I frame no hypotheses. For whatever is not deduc'd from the phænomena, is to be called an hypothesis; and hypotheses, whether metaphysical or physical, whether of occult qualities or mechanical, have no place in experimental philosophy. In this philosophy particular propositions are inferr'd from the phænomena, and afterwards render'd general by induction" (Newton 1934, 547). Hume echoes this rule of experimental philosophy in his description of associations: "here is a kind of

ATTRACTION which in the mental world will be found to have as extraordinary effects as in the natural, and to shew itself in as many and as various forms. Its effects are every where conspicuous; but as to its causes, they are mostly unknown, and must be resolv'd into *original* qualities of human nature, which I pretend not to explain. Nothing is more requisite for a true philosopher, than to restrain the intemperate desire of searching into causes, and having establish'd any doctrine upon a sufficient number of experiments, rest contented with that, when he sees a farther examination would lead him into obscure and uncertain speculations." (T 12-13).

Newton did not, of course, deny that science works with hypotheses; but hypotheses must not rest on metaphysical assumptions.[2] Newton argues that "it is not the Business of Experimental Philosophy to teach the Causes of things any further than they can be proved by Experiments. We are not to fill this Philosophy with Opinions which cannot be proved by Phænomena. In this Philosophy Hypotheses have no place, unless as Conjectures or Questions proposed to be examined by Experiments" (Newton 2002).[3] Newton was accused by Leibniz of using metaphysics, Leibniz claiming it was not possible to explain in any other way how gravitation operates at a distance. This issue was addressed in the Leibniz-Clarke correspondence of 1715–1716, in which Clarke speaks for Newton. The explanation of gravitation entails, according to Leibniz, a reliance on miracles or some occult qualities (that Newton explicitly discards from experimental science). Leibniz argues that it could be perhaps "some immaterial Substances, or some spiritual Rays, or some Accident without a Substance, or some kind of *Species Intentionalis*, or some other *I know not what*, the *Means* by which this is pretended to be performed … And if it is not miraculous, it is false. 'Tis a Chimerical Thing, a Scholastick *occult Quality*" (Leibniz 1717). Clarke, however, defends the empirical grounds of Newton's science and insists that human understanding cannot (and need not) decode the ultimate secrets of Nature. He says that gravitation "is nothing but a *Phænomenon*, or *actual Matter of Fact*, found by *Experience*". Clarke admits that gravitation must have some cause, but until it can be discovered empirically "the Cause, is [therefore] the *Effect* it self, the *Phænomenon*, or the *Matter of Fact discovered by Experience*" (Clarke 1717). Newton seems to consistently defend the view that gravity is a universal quality of bodies, but denies that he is asserting that it is an essential quality.

There have been ongoing debates about the influence of Newton on Hume, instigated by Peter Jones, and analysing the extent of Hume's knowledge of science. Jones argues that there is no evidence that Hume ever seriously studied science; by contrast he showed "a total lack of interest in contemporary science" (Jones 1982, 17); his "fundamentally humanistic orientation separates him completely from any Newtonian influence" (Jones 1982, 13–14). A few years later Michael Barfoot did a

---

[2] The metaphysical aspects of Newton's *General Scholium* are discussed in 4.8.

[3] This passage is taken from Newton's *Account of the Book entituled Commercium Epistolicum Collinii et aliorum de analysi promota*. The *Commercium* contains correspondence relevant to the priority dispute between Newton and Leibniz regarding the invention of Infinitesimal Calculus. It appeared anonymously, but is known to have been written by Sir Isaac Newton.

thorough archival research into Hume's education and discovered that during his studies at the Edinburgh University Hume was a member of the Physiological Library founded by Robert Steuart, and that he participated in his science course; however, Hume was then only 13 or 14 years old.[4] The evidence shows Hume also studied Boyle and became *reasonably* familiar with Newton's *Opticks*, yet his knowledge of science was that of a man of letters predominantly interested in issues other than science. Hume certainly could not comprehend Newton's physics and mathematics. Eugene Sapadin points out that Hume's knowledge of Newton was probably derived from a number of popularizations – for instance from John Keill's lectures published as *Introduction to Newtonianism* (1702) or from Colin MacLaurin's *An Account of Sir Isaac Newton's Philosophical Discourses* (1748).

This gave Hume enough knowledge to appreciate Newton's achievements and, most importantly, his methodology, which was especially emphasized in popular accounts. As Eugene Sapadin correctly argues, "not studying Newton is not the same as not being influenced by him" (Sapadin 1997, 339). Hume laid out the methodological basis of his philosophy in the epistemological parts of his *Treatise* and *Enquiry* and considered it a new logic for moral sciences; or, as we would say today, a methodology for the social sciences. Being in a broad sense a *Newtonian* meant entering in *a new scene of thought*. It is easy today to forget how revolutionary was this change; Gassendi and his circle and Newton's immediate predecessors and contemporaries, such as Boyle, were still making a great effort to liberate natural philosophy from the Aristotelian-scholastic tradition, and despite a vehement criticism of Descartes from the position of the empiricists Descartes' contribution to freeing science from the traditional metaphysical framework was widely appreciated. Also the idea of transferring the Newtonian approach of the study of nature to the human mind and human sciences was not uncommon; in this context, Passmore refers to several verses from Pope's *Essay on Man*: "… all knowledge is, Ourselves to know…Could he [Newton], whose rules the rapid comet bind, Describe or fix one movement of his mind?" (Passmore 1952, 5–6).

Eric Schliesser comes with a different, provocative claim: that Hume intended to 'attack' Newton's authority in natural science in order to secure an independent place for moral philosophy. In Schliesser's view, Newton 'usurped' not only all the fame but asserted that natural philosophy was the foundation for all other sciences, whilst Hume wanted to assert the supremacy of the science of man (Schliesser 2009, 2011; Janiak and Schliesser 2012); he thus suggests that there was a 'territorial conflict' (or a clash of interests) between Hume and Newton. Schliesser even claims that Newton's achievements were not, in Hume's eyes, useful to the public – at least not as useful as Hume's own science of man. This argument seems farfetched. Hume appreciated Newton's method and presumed that by its application to human sciences this new experimental approach would be *enriched*, bringing new much needed discoveries about human nature. In this sense, Hume saw his project as

[4] Hume entered the University at the age of eleven, together with his brother, aged thirteen; it was common then to begin studies at 13 and to leave after 4 years – the first 2 years were devoted to Classics, followed by 2 years of natural philosophy. Hume left the University without a degree.

complementary to Newton's. However, it is true that Hume – at least in the *Treatise* – considered the science of man foundational because the analysis of the operations of the mind formed the foundations for any inquiry; every inquiry is dependent upon how we think. But even then, there is no danger of any territorial conflict between the science of nature and the science of man.

As Hume's foundational claim goes, he says that he proposed, "a compleat system of the sciences, built on the foundation almost entirely new, and the only one upon which they can stand with any security … [this system] is the only solid foundation for the other sciences" (T xvi). A few pages later Hume gives a detailed explanation: "'Tis evident, that all the sciences have a relation, greater or less, to human nature; and that however wide any of them may seem to run from it, they still return back by one passage or another. Even Mathematics, Natural Philosophy, and Natural Religion, are in some measure dependent on the science of Man; since *they lie under the cognizance of men* [emphasis mine], and are judged of by their powers and faculties" (T xv). This foundational role is given to epistemology since it provides the mental map and supplies new logic for moral and even for natural science. One problem, though, remains. In the process of developing his science of man, Hume was entrapped in scepticism toward the legitimacy of the basic concepts we use in thinking and judgment; experimental natural science, working with universal laws, real external bodies and their causal relations, thus becomes undermined in its essence.[5]

2.3 The Copy Principle

The copy principle entails that "we have no idea of any quality in an object, which does not agree to, and may not represent a quality in an impression; and that because all our ideas are deriv'd from our impressions" (T 243). Hume claims that the copy principle is *derived from the observation of our mental mechanisms*; it is introduced as *an empirical generalization* and as a *description* of how we in fact form knowledge. At the same time the copy principle defines the *norm of empiricism* delineating the proper (empirical) domain of our understanding; so far it has a *prescriptive* role.[6] Only knowledge that fulfils this norm can be granted the certificate of being

[5]Mossner points out (and is puzzled by) Hume's rather excessive scepticism concerning natural science; note 4 of the manuscript memoranda on "Natural Philosophy," which seems to belong to his pre-*Treatise* period, reads: "a Proof that natural Philosophy has no Truth in it, is, that it has only succeeded in things remote, as the heavenly Bodys, or minute as Light" (Mossner 1970, 75).

[6]Jerry Fodor emphasized this clash in Hume's copy principle claiming that it contains both an explanation of how ideas acquire their content and a condition of legitimating for ideas. Thus Hume's psychology interferes with his epistemology. Fodor's advice is to focus primarily on Hume's science of mind by the way of "abstracting from the aspects of Hume's theory of mind that are dictated primarily by his epistemology" (Fodor 2003, 33). Similar tendencies appear in Garrett (2011) and Biro (2009). The interpretations of Hume that underrate the clash between the psychological and the normative aspects of the copy principle also tend to underrate the force of his

rationally justified and can receive scientific status. If the copy principle worked like that, Hume's epistemology would be perfect in every way – it would capture the nature of empiricism and serve as a guide for the methodology of experimental science and philosophy. If, as Hume says, all "ideas are preceded by other more lively perceptions, from which they are derived, and which they represent" then Hume could "hope this clear stating of the question will remove all disputes concerning it…" (T 7). However, if we find that there are ideas without corresponding impressions, the whole structure of the copy principle collapses; and that is exactly what happened to Hume when he examined more carefully how we form knowledge. He had to admit that *our basic inferences and judgments violate the copy principle*. He was thus left with three options – to drop the copy principle, to revise it or to admit it was dogmatic. He chose none of them.

The most quoted counter-example to the copy principle is very specific: it refers to the "missing shade of blue" in the whole spectrum of shades of blue (T 5-6); Hume suggests that in this particular case we can supply the sense deficiency with our imagination. He concludes the passage with a verdict many find surprising and which Flew deemed scandalous and outrageous; according to Hume this is "a proof that the simple ideas are not always … derived from corresponding impressions; though this instance is so singular that it is scarcely worth our observing" (E 21, T 6). However, the copy principle is not only a *psychological* generalization based on the observation that "ideas and impressions *appear* [emphasis mine] always to correspond to each other" (T 3), in which case we may allow for a few minor deviants. Hume himself insists that "it still remains true, that all our simple ideas proceed either mediately or immediately, from their corresponding impressions" and concludes that "this is then the first principle I establish in the science of human nature" (T 7). The copy principle is a universal rule and the foundation of the experimental method – yet "any universal generalization is decisively falsified by even one single genuine counter-example" (Flew 1986, 21). Prior to the discussion on the missing shade of blue Hume seemed to agree with this maxim. He invites anybody who doubts the *universal validity* of this rule to challenge it by finding one case in which an idea is not derived from an impression; "it will then be incumbent on us, if we would maintain our doctrine, to produce the impression, or lively perception, which corresponds to it" (E 19-20); if the search is positive the copy principle should be refuted. Yet when Hume comes to discuss the example of the missing shade he ignores this resolution. This seems a considerable inconsistency.

Some philosophers, though, do not judge Hume so harshly. Garrett, for instance, claims that the copy principle is an empirical generalization (that leads to 'conceptual

epistemological scepticism. John Biro, who attributes to Hume an original contribution to the 'New science of the mind', argues that "his skepticism is better understood as one about pretended supra-scientific metaphysical knowledge, rather than about scientific knowledge itself" (Biro 2009, 46). But Hume in his epistemological writings was deeply sceptical about all empirical knowledge including science. The foundational aspect of epistemology was constitutive for the philosophical discourse at that time, and Hume was no exception. His provocative solution to the legitimation crisis lies in a carefree ('Pyrrhonian') abandonment of the problem and in shifting the inquiries to moral science instead.

empiricism') and that it only aims to show that experience is the basic (but not the absolutely necessary) condition for understanding. Garret admits that Hume's confidence in the copy principle is not based on extensive empirical research, but that "[Hume] thinks that he has observed enough about human nature to know that, if counter-examples to the copy principle were at all common, then he would have heard of them" (Garrett 1997, 48). According to Garrett, a few exceptions do not change the overall empiricist grounding of knowledge-formation; after all, we are continually observing how our minds work and adding new facts to the initial picture. Garrett concludes that the copy principle "provided Hume a motivation to form more detailed investigations into the cognitive processes underlying the use of central yet problematic concepts" and "to focus on those investigations in the area of experience and practice, on which these concepts are based" (Garrett 1997, 57). Garrett is right about Hume's aim to 'anatomize' our nature, here specifically the mechanism of our mental life, but he underestimates the normative impact of the copy principle. Generally, his approach to Hume seems too conciliatory; he suggests that ideas lacking their corresponding impressions may be rationally justified, at least to a degree, since they are also products of our reasoning – though not certified by this reasoning. He is rather evasive about these contradictions in Hume's philosophy and following his account it may be hard to understand why Hume was a sceptic; it would be also difficult to understand why the awareness of sceptical contradictions should cause such suffering and anxiety to him, as described in the Conclusion of Book I of the *Treatise*, or why he consistently admitted the hopelessness of efforts to find a solution to the sceptical crisis.

To sum up, Hume's scepticism unfolds from the *contradiction between the norm of empiricism (defined in the copy principle) and the real processes in our thinking*. These processes do not follow the steps prescribed by the copy principle and transcend the narrow, straightjacketing scope delineated by Hume's strict empiricism. This sceptical conclusion concerns the basic and most general inferences and judgments; the causal relations and the 'objectivizing' nature of our thought. Causation is for Hume the constitutive inference for our comprehension of the world. As he says, "if we anatomize all the other reasonings of this nature [concerning matter of fact], we shall find, that they are founded on the relation of cause and effect" (E 27). We expect that fire always produces heat or that bread always nourishes – the former is a cause that necessarily produces that latter effect. However, if we limit our judgment to what we directly observe we have no evidence of any causal relation. We only observe two distinct phenomena or events; we also have the experience that they always occur together, in succession. We have experience of their *constant conjunction* but not of their *necessary connection*; it is illegitimate to extrapolate from the past to a universal necessity that will be valid in the future.

2.4 Ideas Without Impressions?

Hume exposes this contradiction to broad daylight. With regard to causation he acknowledges that we observe only single, loose events of which one has always preceded the other. However, the idea of causation entails necessary connection of two events and allows no exception; there must be something – some energy, power or efficacy – in the cause that always *produces* the same effect. Therefore, as Hume says, "we must find some impression, that gives rise to this idea of necessity, if we assert we have really such an idea"; in fact, though, we "immediately perceive, that they ['cause' and 'effect'] are *contiguous* in time and place, and that the object we call cause *precedes* the other we call effect. In no one instance can I go any farther, nor is it possible for me to discover any third relation betwixt these objects" (T 155-6). The conclusion that "no objects have any discoverable connexion together, and that all the inferences, which we can draw from one to another, are founded merely on our experience of their constant and regular conjunction" (E 111-12) implies that, in accordance with the copy principle, we should never even have the idea of a necessary connection – and thus no idea of causation. And this is exactly what Hume claims. "We have no idea of this connexion; nor even any distinct notion what it is we desire to know, when we endeavour at a conception of it" (E 77). In one of his most extreme statements Hume even admits that this implication of empiricism may paralyze our lives; we may observe, he says, that "the necessary conclusion *seems* to be, that we have no idea of connexion or power at all, and that these words are absolutely without any meaning, when employed either in philosophical reasonings, or common life" (E 74). If applied to common life, we not only could not think and judge but, taken to the extreme, we could not live. However, Hume sees the absurdity of such a conclusion which is contrary to all evidence. He thus corrects the argument concerning the relation of cause and effect: "in all these expressions *so apply'd*, we have really no distinct meaning, and make use of common words, without any clear and determinate ideas" (T 162).

At this stage, Hume introduces another power – it is human nature endowed with instincts, including cognitive instincts, and common sense. Nature is like a sweeping current that wafts away all our doubts concerning the idea of causation and external existence and makes us ignore the copy principle. Due to our nature we believe in causal power and in the independent existence of things around us; this belief is instant, irresistible and involuntary, allowing no discussion. When we *reflect* upon the situation we must conclude, under critical scrutiny, that we cannot have any ideas of necessity, or external existence, since they cannot be derived from experience. Yet we *have them*, we have the thoughts and the words – and we call them 'ideas' because it is natural to do so despite the fact that they are neither proper nor adequate. Therefore, Hume's declaration in the *Abstract* that "if no impression can be produced, he [i.e. Hume himself] concludes that the term is altogether insignificant" cannot be interpreted literarily (T *Abstract* 649). The idea of necessary connection *governs our thought and actions but is rationally illegitimate*.

Hume remains true to the Newtonian experimental method and rejects any explanations of causation using references to some hidden mechanisms by which the "human mind is actuated in its operations" (E 14); therefore, "the ultimate springs and principles are totally shut up from human curiosity and enquiry" (E 30). Sceptical research into causation (or existence) is irresolvable and epistemology must move to different themes. Hume thus shifts his attention onto *how these ideas arise in our mind*. He introduces new terms that are formative of this process – custom, imagination and belief. The construction of the 'inadequate ideas' begins with a *repeated* experience of two events being conjoined. If Adam "created in the full vigour of understanding, without experience" observed billiards for the first time he would not be able to predict the movement of a ball struck by another ball (T *Abstract* 650). In a similar way, Adam "could not have inferred from the fluidity, and transparency of water, that it would suffocate him, or from the light and warmth of fire, that it would consume him. No object ever discovers, by the qualities which appear to the senses, either the causes which produced it, or the effects which will arise from it" (E 27). But after he has repeatedly observed billiards being played he becomes *accustomed* to the conjunction of the two events and he would then "always conclude without hesitation, that the second [ball] would acquire motion" (T *Abstract* 651).

Thus custom stimulates our imagination and forms a belief in the necessity of the connection between the cause and the effect even though it does not have rational support. Repeated experience of the constant conjunction of events "immediately, by the force of custom, carries the imagination to conceive that object, which is usually conjoined to it; and this conception is attended with a feeling or sentiment, different from the loose reveries of the fancy. In this consists the whole nature of belief" (E 48). As Hume describes in his second definition of cause, "*the appearance of one object always conveys the thought to that other*" (E 77); it is a kind of conditioned reflex in which "the mind anticipates the senses" (E 77) or, we could say, it jumps ahead of the senses and 'cheats' the copy principle. The same pattern is applied in the case of our ideas regarding the external existence of things like a tree or a table. In a philosophical reflection we perceive fleeting and interrupted impression; they come and go, mix and change. In the empiricist view we should not be able to form the idea of a thing that remains unchanged (or continues to exist) when we turn our eyes away; we have no impression of a distinct and constant existence. Despite that, due to our natural propensity, we "unite these broken appearances by the fiction of a continu'd existence" (T 204).

On the one hand our imagination produces a belief in the identity of resembling perceptions and on the other hand we know that perceptions are unstable, always in flux. The first view is typical of the vulgar view common to, as Hume says, "all the unthinking and unphilosophical part of mankind" (T 204). Opposed to this is the philosophical view, which uses critical reflection of our mental operations; in this view, the belief is unfounded since we only perceive single unconnected sense data. However, even a philosopher who is aware of this dilemma cannot resist the belief. He, at times, philosophizes in a closet, isolated from common life and haunted by scepticism; but he also finds himself "absolutely and necessarily determin'd to live, and talk, and act like other people in the common affairs of life" (T 269), and then all doubts vanish like a puff of smoke. Being both a sceptical philosopher and a common man he is trapped in an unenviable position, torn between belief and reason and being perplexed (and traumatized) by this contradiction. This contradiction between reason, that performs *profound reflection* and finally discovers "surprising ignorance and weakness of the understanding" (E 76), and nature, that commands *blind submission* to beliefs, is the leading theme in Hume's epistemology.

2.5 The Constitution of Our World

Following up on the previous chapter on the Cartesian strands of Hume's scepticism it is important to stress that on the empiricist level his scepticism – resulting from the contradiction between empiricist norms and the natural cognitive processes that violate these norms – *does not have ontological relevance*. All Hume's considerations of causation or existence concern the realm of *the activities of our hearts and minds*. Since Descartes proclaimed that we were aware only of our own mental contents the constitution of objectivity became a primary philosophical concern. Hume looks at *how our thinking constitutes objectivity* – it is thus objectivity generated by the subject from his own resources and by his own dispositions (associations, imagination, sentiments, and beliefs. Our knowledge is about entities produced by our minds: Hume concludes that causation "is a quality which can only belong to our mind" (T 168) and that two [phenomenal] objects "have acquired a connection in our thought" (E 76); analogically, when Hume talks about external existence he investigates how *we come to have the idea* of an object and *why we believe* that, for example, this table has a real existence. This level of Hume's scepticism thus concerns *the world of our thought* and brings forward the importance of the activity of the self, the constitution of objectivity *by the subject*.

Descartes merely acknowledges it is a disposition of the human mind: "there can be no ideas that are not as it were (*tanquam*) ideas of realities" (Descartes 1979[1641], 84). Hume also takes for a fact that "whatever we conceive, we conceive as existent" (T 67) but pays more attention to the mechanisms of this constitution; he works on our mental geography, the anatomy of human nature, of which thinking is only one part. This aspect of Hume's scepticism strongly influenced Kant. The fact that we noetically make our world awoke him from his "dogmatic slumber"; and Hume's account of these operations of the mind made Kant look for a sounder foundation than Hume's purely empiricist suggestions. Kant's awakening is commonly associated with his criticism of Hume's psychological account of causation – but that is only one part of the story. Kant's awakening entailed a break away from the metaphysical rationalism of Christian Wolff, the leading Enlightenment philosopher in Germany before him. In the "Preface" to the second edition of his *Critique of Pure Reason* Kant calls him the greatest of dogmatic philosophers; although he appreciates Wolff's contribution to establishing a deductive mathematical method applicable to both philosophy and science, creating thus a modern (German) philosophical vocabulary, he abandons Wolff's metaphysical anchor of philosophy – for Wolff *philosophia rationalis* is framed in theology.

In this sense, the abandonment of metaphysics led Kant to his famous Copernican turn. As Copernicus denied that celestial bodies revolve around the stationary Earth, but declared that the Earth itself moves, Kant replaced the assumption that "all our cognition must conform to objects" with the opposite, namely that "objects must conform to our cognition" (Kant 1996, B xvi, 57). The first impulse, given by Descartes and further developed by Hume, was finally completed by Kant. Kant criticizes Hume for his purely empiricist and psychological explanation of cognition – we cannot explain from experience alone the universal necessity entailed in our thinking and knowledge. Kant's plan was to unite the two formative elements of knowledge, experience and understanding, on the basis that the latter cannot be derived exclusively from the former. Being more 'realist' than Hume Kant claims that perceptions, as the material of experience, are not of "unknown origin" but come from outside, from things in themselves that affect our senses; in this context, I repeat Kant's famous statement: "for what else might rouse our cognitive power to its operation if objects stirring our senses did not do so?" (Kant 1996, B 1, 43).

This may seem a rather dogmatic introductory sentence for somebody who has just awakened from a dogmatic slumber. Even if the intention was to avoid the dissolution of all experience in pure subjectivity the claim would remain dogmatic. Why, then, does Kant need the thing in itself? The explanation is given only later in the *Critique*; Kant argues that the existence of the thing in itself is a precondition of our ability to conceive of the permanence and continuance of perceptions in time. This awareness cannot, according to Kant, be deducible from the transcendental activity of the self since the awareness of the permanence of the self is itself a precondition of this kind of perception. Kant says: "I am conscious of my existence as determined in time. All time determination presupposes something *permanent* in perception. But this permanent something cannot be something within me, precisely because my existence can be determined in time only by this permanent something … i.e. the consciousness of my existence is simultaneously a direct consciousness of the existence of other things outside me" (Kant 1996, B 276, 290). This is yet another issue relevant to Hume.

Kant's primary target is Hume's conception of experience. For Kant, experience itself is not 'given' in perceptions as Hume assumes; "even though all our cognition starts **with** experience [raw material of sense impressions], that does not mean that all of it arises **from** experience" (Kant 1996, B 1, 44). Experience is dependent on pure intuition (time and space) and understanding is dependent on pure concepts of understanding (categories); both are formal *a priori* dispositions. Experience presupposes a *form* of ordering of the matter for cognition, taken from the senses. What we are directly conscious of are already perceptions *experienced in time and space*, later submitted to the categorial synthesis through which they derive their *objective meaning*; these are, as Kant says, the conditions of the possibility of experience which are, at the same time, the *conditions of the possibility of the objects of experience*. Hume's position, in which knowledge is formed from impressions by the mechanisms of association that are themselves triggered by this sense experience, is thus unacceptable to Kant; Hume looks like a Don Quixote attacking windmills in the vain hope of producing sound epistemology from *mere* observation.

In the *Prolegomena* Kant criticizes Hume's empirical derivation of causal necessity from the perception of one appearance constantly followed by another; in Kant's famous metaphor, Hume, owing to his empiricism, "deposited his ship [epistemology] on the beach (of skepticism) for safekeeping,where it could then lie and rot" (Kant 2004, 11–12). Although it is believed that Kant read only Hume's *Enquiry*, this formulation seems to refer to Hume's sceptical conclusions from the *Treatise* concerning the sorry state of our understanding wherein he compares reason to the "weather-beaten vessel" (T 263). Kant wants to continue Hume's voyage by supplying the ship with a "pilot [reason], who, provided with complete [*a priori*] sea-charts and a compass, might safely navigate the ship … following sound principles … drawn from a knowledge of the globe" (Kant 2004, 12) – in other words, equipped with the *a priori* dispositions that guarantee the necessary unity of experience and knowledge. Hume's epistemology cannot, according to Kant, accommodate reason and be the foundation of all other sciences, as Hume proposed in the *Treatise*. Kant was also an admirer of Newton and even claimed that it was absurd to hope that maybe another Newton might some day arise. He perceived his epistemology as the right foundation of Newton's natural science; that plus Kant's own moral science – "the starry sky above me and the moral law within me" – were for Kant the two things that "fill the mind with ever new and increasing admiration and reverence" (Kant 2002, 203). He hoped to deliver what he felt Hume had failed to provide.

Kant's criticism of Hume is often considered too sceptical and too rationalistic; not adequately appreciative of Hume's naturalism and dwelling on the *a priori* ('innate') basis of our experience. In this view, Kant represents a continuation of the tradition of "rationalistic metaphysics" of the Cartesian provenance whose "consummate defeat" is the main achievement of Hume's philosophy (Millican 2006, 50–1). Hume obviously rejected Descartes' innate ideas on the basis of his empiricism; however, Hume explicitly declared that the main problem of his epistemology – and the main source of his scepticism – lies elsewhere, in the impossibility of providing a rational warrant for the natural production of beliefs. Hume also recognizes that this scepticism is a defeat of the rationality of knowledge. However, Kant's criticism reveals an even *deeper level of scepticism* that results from pure empiricism, a level of which Hume was unaware. Kant claims that *Hume's empiricism fails to provide the philosophical foundations for the mere possibility of experience*; thus the main problem of Hume's philosophy is (for Kant) not the problem of the justification of beliefs but the problem of the alleged natural production of beliefs. According to Kant, Hume did not acknowledge that our 'natural dispositions' must be *formally* prior to experience and not triggered by repeated experience. One may disagree with Kant but it is a valid argument that moves Hume's scepticism into a new domain.

This issue of learning and knowledge-building is high on the agenda of contemporary cognitive science and it seems that Kant was ahead of his time in opening up the problem of innate dispositions. Are our cognitive abilities, either generally or some specific dispositions such as the linguistic, mathematical and musical – innate, that is genetically endowed, or are they acquired by learning through the evolution of our species and through interaction with our environment? Or can there be some overlap, for instance when long-term adaptation strategies result in genetic changes and become innate? The debate is far from settled and new medical and computing techniques are enabling us to penetrate ever deeper into these problems. Hume (with Locke) and Kant can be seen as the first philosophers to embark on the project of constructing a science of the mind. They are also the founders of the two opposing trends which continue today in the empiricist/nativist debate; Hume and Kant anticipated the debate but lacked many of the experimental and research opportunities which twenty-first century science has at its disposal.

However, it would be wrong to reduce Hume's philosophical importance to cognitive psychology and to leave out the sceptical issue of the rational warrant of epistemology. The psychological interpretations of Hume (e.g. Garrett 1997; Fodor 2003; Biro 2004) show Hume's relevance to modern cognitive science, and to some extent link his naturalism to Quine's project of naturalized epistemology. As emphasized especially by Garrett, Hume does not ask about the normative foundation of beliefs but seeks to explain the natural causes of them; Garrett concludes that according to Hume our reasoning pertaining to matters of fact (our inductive reasoning) is not affected by any higher-order rational inference though it involves the exercise of reason: "[Hume] is denying only that we come to engage in this species of reasoning as a result of any piece of reasoning about it" (Garrett 1997, 94). In a detailed criticism of Garrett, Millican argues that Hume (especially in the *Enquiry*) no longer uses the term 'reason' in a psychologistic way and that his principal focus is on epistemology and the questions of *rational warrant* (Millican 1998, 141–160). Hume, as mentioned above, admits defeat on this point. Scepticism with regard to the rational foundations of knowledge is inescapable and psychologically traumatic. Therefore, he quickly moves on to more positive domains of inquiry, including our natural disposition to acquire knowledge.

Yet in Hume's philosophy this positive domain is not consistently united with his sceptical conclusions. On the one hand, Hume is totally sceptical regarding the rational foundations of knowledge; and on the other hand he is totally non-sceptical about the power of nature in forming knowledge. Millican describes this as a *false dichotomy* due to the classic sceptical *caricature* of Hume's philosophy (Millican 2007, 194–197). He illustrates this "false indiscriminating scepticism" with Stroud's "delightful phrase": "as far as the competition for degrees of reasonableness is concerned, all possible beliefs about the unobserved are tied for the last place" (Stroud 1977, 54). Millican argues that these two attitudes, the sceptical and the non-sceptical, are mutually supportive and not exclusive since we are psychologically unable to refrain from forming beliefs about the unobserved. Yet does the fact that we do indeed both form beliefs and discredit their rational basis demonstrate the unity of these two attitudes? Furthermore, would any classic sceptical Humean deny the coexistence – though not necessarily harmonious – of these attitudes? Stroud himself claims that Hume's scepticism was paving the way for his naturalistic account of human nature. Richard Popkin and, to an extent, Robert Fogelin also agree that Hume's naturalism and his scepticism coexist – they are neither at war nor in partnership; Hume's scepticism remains theoretically totally unmitigated and his naturalism remains totally rationally unjustified, despite being pragmatically functional since our natural inclinations ignore the theoretical condemnation. The relationship between the two attitudes can vary depending upon the angle from which we view them; pragmatically and practically they are supportive, if that is what Millican claims, whilst rationally they are exclusive, and the natural attitude does not require any theoretical support since it represents an instinctive embracement of the "bare necessities of life".[7] Hume's ambivalence to these persisting tensions between the various perspectives on reason, nature and knowledge is linked to his Pyrrhonism, which instructs us to be tolerant and even indifferent to contradictions, as discussed in Chap. 5, despite such an injunction being unacceptable to the scientific, analytical discourse prevailing in contemporary interpretations of Hume.

Millican's main aim is to show that Hume's scepticism does not undermine the possibility of inductive science. Of course it does not – provided we accept that science is founded on instincts, custom and belief, justified by its own efficiency and results. Hume never denied that we use (rationally unwarranted) inductive reasoning in forming our knowledge; according to him – but denied by Kant – this is what we naturally do. However, as argued later in this book, Hume never considered this self-justifying *practice* a solution to the lack of *rational justification*. And since no philosophical solution was available, Hume moved away from epistemology into moral science. He was not too disconcerted by the fact that science was founded on animal instincts since *natural science did not interest him*; the Scottish Enlightenment was oriented more to humanism in a broad sense than to natural science. Men of letters from Glasgow, Aberdeen and Edinburgh formed groups and societies with a wide range of overlapping interests – for example moral philosophy (Francis Hutcheson), economy, history and social philosophy (Lord Kames, Adam Smith, Adam Ferguson), literature, linguistic and poetry (Allan Ramsay, James Burnett, Hugh Blair, Alison Rutherford), the philosophy of common sense (George Campbell, Thomas Reid, James Beattie); Hume was a member of "The Select Society" – later the "Poker Club" – a group of *literati* which met in Edinburgh.

He was concerned for the well-being of mankind but did not think it could be achieved purely or even primarily through progress in natural science; besides, he believed that everything in this field had been already discovered by Newton. Therefore Millican's arguments that Hume was a fervent advocate of inductive science, or that Hume was both "deeply sceptical about induction (in a sense), *and* totally committed to inductive science" (Millican 2007, 195) seem farfetched. Hume was by no means an advocate of natural science, let alone a fervent one – he simply did not care about science (see Sect. 2.2). Rather than advocating inductive science, Hume backed out of this field; and in a loosely Pyrrhonian way he accepted that unlike beasts we are capable of profound reflection and like beasts we follow

[7] The Baloo song from the movie *The Jungle Book* expresses this attitude quite well: "Look for the bare necessities/The simple bare necessities/Forget about your worries and your strife/I mean the bare necessities/Old Mother Nature's recipes/That bring the bare necessities of life."

nature. The natural instincts of human beings include cognitive instincts, linked to other unique faculties such as abstraction, language and memory – but this uniqueness does not in principle modify the basic duality inherent in the human constitution, a duality which cannot be unified by any metatheory (not even by mitigated scepticism, as discussed in Chap. 5).

References

Biro, J. 2004. Cognitive science and David Hume's science of the mind. In *Truth, rationality, cognition, and music*, ed. K. Korta and J.M. Larrazabal, 1–20. Dordrecht: Springer.

Biro, J. 2009. Hume's new science of the mind. In ed. F.D. Norton and J. Taylor, 40–70.

Blackburn, S. 2008. *Hume*. London: Granta Books.

Clarke, S. 1717. *Leibniz-Clarke correspondence of 1715–1716*. Clarke's fifth reply. London. http://www.newtonproject.sussex.ac.uk/view/texts/normalized/THEM00235. Accessed 5 Mar 2015.

Descartes, R. 1979 [1641]. Meditations on first philosophy. In *Descartes' philosophical writings*. Trans. E. Anscombe and P.T. Geach, 59–125. Nelson's University Paperbacks.

Flew, A. 1986. *David Hume. Philosopher of moral science*. London/New York: Basil Blackwell.

Fodor, J.A. 2003. *Hume variations*. Oxford: Clarendon.

Garrett, D. 1997. *Cognition and commitment in Hume's philosophy*. Oxford: Oxford University Press.

Garrett, D. 2011. Hume's theory of ideas. In ed. E. Radcliffe, 41–58.

Hume, D. 2005 [1761]. *The history of England, from the invasion of Julius Cæsar to the abdication of James the Second, 1688*. Vol. VI. Michigan: Michigan University Library.

Janiak, A., and E. Schliesser (eds.). 2012. *Interpreting Newton: Critical essays*. Cambridge: Cambridge University Press.

Jones, P. 1982. *Hume's sentiments: Their Ciceronian and French context*. Edinburgh: University of Edinburgh Press.

Kant, I. 1996 [1781]. *The critique of pure reason*. Trans. W.S. Pluhar. Indianapolis: Hackett Publishing Company.

Kant, I. 2002 [1788]. *The critique of practical reason*. Trans. W.S. Pluhar. Indianapolis: Hackett Publishing Company.

Kant, I. 2004 [1783]. *Prolegomena to any future metaphysics: That will be able to come forward as science*. Trans. G. Hatfield. Cambridge: Cambridge University Press.

Leibniz, G.W. 1717. *Leibniz-Clarke correspondence of 1715–1716*. Leibniz' fifth paper. London. http://www.newtonproject.sussex.ac.uk/view/texts/normalized/THEM00234 Accessed 5 Mar 2015.

Locke, J. 2011 [1689]. *An essay concerning human understanding*, ed. P.H. Nidditch. Oxford: Oxford University Press.

Millican, P. 1998. Hume on reason and induction: Epistemology or cognitive science. *Hume Studies* Vol. XXIV:1, 141–160.

Millican, P. 2006. The context, aims and structure of Hume's first Enquiry. In *Reading Hume on human understanding*, ed. P. Millican, 27–67. Oxford: Oxford University Press.

Millican, P. 2007. Humes old and new: Four fashionable falsehoods, and one unfashionable truth. *Proceedings of the Aristotelian Society Supplementary Volume LXXXI*, 163–199.

Mossner, E.C. 1970. *The life of David Hume*. Oxford: Clarendon.

Newton, I. 1934 [1713]. *Mathematical principles of natural philosophy*. Trans. A. Motte, ed. F. Cajori. Berkeley: University of California Press.

Newton, I. 2002 [1714]. *An account of the book entituled Commercium epistolicum collinii et aliorum, de analysi*, ed. D.R. Wilkins. http://www.maths.tcd.ie/pub/HistMath/People/Newton/CommerciumAccount/CommAcc.pdf. Accessed 20 Jan 2016.

Norton, F.D., and J. Taylor (eds.). 2009. *The Cambridge companion to Hume*. Cambridge: Cambridge University Press.

Owen, D. 2009. Hume and the mechanics of mind: impressions, ideas, and associations. In ed. F.D. Norton and J. Taylor, 70–105.

Passmore, J.A. 1952. *Hume's intentions*. Cambridge: Cambridge University Press.

Radcliffe, E. (ed.). 2011. *A companion to Hume*. Malden: Wiley-Blackwell.

Sapadin, E. 1997. A note on Newton, Boyle, and Hume's "experimental method". *Hume Studies*, XXIII, 2: 337–344.

Schliesser, E. 2009. Hume's attack on Newton's philosophy. *Enlightenment and Dissent* 25: 167–203.

Schliesser, E. 2011. Newton's challenge to philosophy: A programmatic essay. *Hopos: The Journal of the International Society for the History of Philosophy of Science* 1: 101–128.

Stroud, B. 1977. *Hume*. London: Routledge.

Chapter 3
The Rebirth of Pyrrhonism in Hume's Time (and Before)

Abstract The revival of Pyrrhonism in Western Europe was facilitated by Latin translations of the work of Sextus Empiricus in the sixteenth century, and further promoted by Michel Montaigne who brought scepticism to the forefront of philosophical interest. Ancient Pyrrhonism began to subvert all the established dogmas as a matter of principle and was thus a fuse that accelerated both the decline of scholastics and the formation of the new position based on the confident self. After all, even though scepticism was a destructive method, based on subversive arguments concerning the reliability of our senses and reason, these arguments had their source in man's own ability to think. Pyrrhonian scepticism found a fertile ground in France, in the works of natural philosophers like Pierre Gassendi and Samuel Sorbière and, in the next generation, Pierre-Daniel Huet and Simon Foucher. They accepted the fact that Pyrrhonism could not be defeated and tried to find some operational space for science within its framework by replacing the ideal of certainty of knowledge by probability and in calling for modesty in our knowledge claims. Hume drew on these ideas but proposed a more radical, unmitigated form of scepticism inspired by Bayle.

Keywords Pyrrhonism • Crisis • Mitigated scepticism • New science

3.1 The Crisis

The intellectual environment at the turn of the seventeenth century was affected by a deepening crisis in the Aristotelian-scholastic world view. This crisis had been to a large extent caused by the Renaissance turn to man, to this life and this world. The self-confidence of man in his own abilities was forcing its way through the straightjacket of scholastic dogma, in which natural philosophy was determined by theological teachings; at its most extreme, any hint at other than a finite geocentric picture of the universe would have been considered blasphemy, denying the truth of the Bible (often with fatal consequences for the person holding such views). Similarly, the Aristotelian qualitative conception of nature, with each part fulfilling its inner purpose and altogether forming a purposeful whole, presented an obstacle not only to the new conception of matter. On a general level scholastic prevented the

Z. Parusniková, *David Hume, Sceptic*, SpringerBriefs in Philosophy,
DOI 10.1007/978-3-319-43794-1_3

freedom and the right of man to explore the world with an open mind and to let the imagination fly high, to carry out one's own experiments and push back the boundaries of human knowledge independent of any higher authority. This surge permeated the atmosphere of the time, encouraging the rise of knowledge and science as well as the discoveries of new lands and customs, the development of trade and technical innovations (such as in military engineering and accounting). Life was an adventure. Under this force the old dogmas had begun to shake in their foundations.

At this time another factor intensified the crisis. Ancient Pyrrhonism emerged from oblivion and began to subvert all the established truths as a matter of principle, not for their specific content but for the incoherence of any judgements that, when scrutinized by the sceptical method, begin to disintegrate from within. The traditional doctrines were exposed as lacking rational support. The revival of ancient scepticism thus caused a huge turmoil in the theological and natural philosophy of the time. The old was shaking but the new ground was not yet prepared and had no plausible theoretical expression at its disposal. Scepticism was a fuse that accelerated both the decline of scholastics and the formation of the new position based on the confident self. After all, even though scepticism was a destructive method, based on subversive arguments concerning the reliability of our senses and reason, these arguments had their source in man's own ability to think. In this unstable environment various responses to scepticism were clashing and mixing, creating a melting pot of multiform influences and new visions. After centuries of relative calm, philosophers found themselves on the threshold of a new era and searched for a new philosophical grounding.

Amidst all the philosophical proposals of how to tackle the newly risen Pyrrhonian scepticism, two philosophers of two generations proposed solutions of utmost importance which influenced further developments in philosophy. They were Descartes and Hume. In Chap. 1, I emphasized their common ground, consisting in their full awareness of the sceptical implications of the position based on the autonomous mind; our mind operations became the starting point and the limiting horizon of philosophy – this resulted in the 'loss of the world' and phenomenalism. Yet in their solutions to the sceptical crisis, Descartes and Hume were irreconcilable opponents. Descartes was determined to beat scepticism in order to save reason and thus to save natural science. Hume accepted that scepticism was irrefutable by argument but took this fact as an *opportunity to investigate our natural cognitive mechanisms* that are not rationally justifiable but are essential for life. Therefore, Hume's philosophy does not share the usual modern goal of being the foundation of science and thus having a normative epistemological function (relying on the authority of reason) but, in the sphere of epistemology, orients inquiries towards the observation of mental processes as an expression of human nature. Negative findings concerning the power of reason did not bother Hume too much since he was not active in natural science; he also saw that our natural instincts were sufficient for the acquisition of knowledge and for life.

In the broadest sense, Hume overcomes scepticism by a 'side-step' that shifts the focus of philosophy to an area in which scepticism cannot instigate its destructive force – the study of man in all his manifestations, including our understanding, moral and social science. This is not to say, as Kemp Smith does (quoting Hume's *Treatise*), that reason is subordinate to passions (Kemp Smith 2005, 11); the weakness of reason shows the impotence of an epistemology that has the ambition to serve as a foundational discipline. Here, Hume adopts unmitigated epistemological scepticism that is compromised neither by our natural tendencies nor by the moderate scepticism adopted in common life. By giving up the goal of formulating a methodology suitable for science, Hume draws on – and extends – the Pyrrhonian legacy in which philosophy has a practical task to be a guide to happiness; Hume's investigations into our sentiments and imagination, our inter-subjective relations and our involvement in society are meant to fulfil this goal. In this approach to philosophy Hume stood outside the mainstream of philosophers for whom natural science was the primary concern.

3.2 The Revival of Pyrrhonism

The revival of Pyrrhonism in Western Europe was facilitated by two Latin translations of the work of Sextus Empiricus. The first translation of the *Pyrrhoniae Hypotyposes* (*Outlines of Pyrrhonism*) by Henri Estienne appeared in 1562, followed by the translation of *Adversus Mathematicos* (*Against the Mathematicians*) in 1569 by Gentian Hervet. Scepticism fell on fertile soil and soon gained enormous popularity in academic circles. The initiating role in this process was played by Michel Montaigne, who presented and discussed various sceptical views in his essay, *An Apology for Raymond Sebond* (1580); this widely-read essay brought scepticism to the forefront of philosophical interest and "caused a *coup de grâce* to an entire intellectual world" (Popkin 2003, 56). Ancient scepticism to a large extent formed early modern philosophy, including the philosophy of David Hume; Pyrrhonism actually culminated in Hume – and also ended with him. Kant, in opposition to Hume, restored the authority of reason in epistemology fully in accordance with the spirit of the Continental Enlightenment.

The sceptics' road back to fame was long and difficult. Their teachings did not have much influence after the Hellenist period; they were partly recorded in Diogenes Laertius' *Lives and Opinions of Eminent Philosophers* but this text was virtually forgotten, along with Sextus' records of scepticism. Academic scepticism, represented by Cicero, was brought into the limelight via Augustine's polemics with the *Academica* in his *Contra Academicos*; but Cicero was well known throughout the Middle Ages mainly for his literary qualities, for his eloquence and beautiful language and his political role in Roman history. The texts of the Greek sceptics were diffused in Byzantium and were not known in the Latin cultural world – they still had a long way to travel from Constantinople to Italy (in the early fifteenth century) and finally, a century later, across the Alps to Western Europe, specifically to France. This adventurous journey is discussed in detail by C.B. Schmitt (1983) and also by Richard Popkin (2003). Montaigne's appreciation of scepticism was the spark which lit the fire that contributed to the disintegration of scholasticism and to the birth of modern philosophy. Pyrrhonism possibly enjoyed more fame at that time than in the ancient world.

Richard Popkin initiated wide-ranging historical and philosophical research into the revival of Pyrrhonism in Europe. He also initiated a lively debate which resulted in a new and fascinating picture of the philosophical developments in this early modern era. His life-long work is concentrated especially in his two pioneering books – *The History of Scepticism from Savonarola to Bayle* of 1979 (an enlarged version of *The History of Scepticism from Erasmus to Descartes* of 1960), and *The High Road to Pyrrhonism*, 1980. The latter work focuses mostly on Hume. Both books prompted many discussions, some of which are summed up in *The Skeptical Tradition*, edited by Myles Burnyeat in 1983. This volume covers both ancient scepticism itself and its impact on famous philosophers of the early modern era; on Gassendi (R. Walker), Descartes (B. Williams), Locke (M. Brandt Bolton), Berkeley (R. Popkin) and Hume (R. Fogelin). Also important is Terence Penelhum's contribution to this debate adding yet another aspect of the legacy of scepticism, namely the rise of sceptical fideism; his work from the last four decades of the twentieth century is summed up in his Themes in Hume (2000). Further, Fogelin's *Hume's Skepticism in the Treatise of Human Nature* (1985) should be mentioned together with his *Pyrrhonian Reflections on Knowledge and Justification* (1994) and the volume *Pyrrhonian Skepticism* edited by Walter Sinnott-Armstrong (2004).[1] Recent publications that develop sceptical themes in ancient, modern and contemporary contexts include, among others, Giani Paganini (2003), J.R. Maia Neto et al. (2009) and Diego Machuca (2011). They build on Popkin's legacy and with admirable scholarly erudition enlarge the area of investigation of scepticism.

In a broader sense, Popkin appeals to philosophers today not to neglect serious historical research; without that, he says, we cannot properly comprehend the language and semantic they used in addressing the central intellectual issues of the time, the philosophical priorities and goals. Popkin detects a tendency in contemporary philosophy to divide the historical and the 'analytical' with the former often being transferred to the history of ideas, conceived as belonging more to history than to philosophy, a field supposed to be more 'archaeological' than inventive or relevant to present philosophical concerns. I appreciate Popkin's appeal for this historical contextual approach and agree with him that "if one replies that the sort of activity … is a waste of time, I can only reply that it is necessary if one is going to base one's claims on accurate rather than inaccurate information" (Popkin 1993, 5). This applies especially strongly to Hume, who lived and worked in the time of turmoil in which new ways of thinking were crystallizing from the decomposing

[1] In the Introduction, Sinnot-Armstrong surprisingly does not credit Popkin with opening this Pyrrhonian line of interpretation but claims that "this tradition has been revived and extended recently in a major work by Robert Fogelin" in his *Pyrrhonian Reflections on Knowledge and. Justification* (Sinnott-Armstrong 2004, 4). However, Fogelin participated in the debates that drew on Popkin, specifically in the volume edited by Burnyeat (1983) and surely exploited these sources in developing his own version of the Pyrrhonian influence on modern philosophy (with an emphasis on Hume).

metaphysical systems of the past (hence viewed as a 'dark' age). Both Locke's and Hume's scepticism concerning knowledge were strongly influenced by the debates that took place in France and were formed by Descartes, Gassendi, Huet and Bayle. This context offers a deeper and richer picture of, in our case, Hume's philosophy.

3.3 Montaigne and the Expansion of Pyrrhonism

Due to the translations of Sextus Empiricus, Pyrrhonian sceptics could be read in Latin; however, through Montaigne they spoke in French to a wider public. In his *Apology* he discusses the arguments of Sextus and Diogenes, Cicero and Erasmus – and compares them with various anti-sceptical positions (Plutarch, Lucretius and Epicurus). The *Apology* was designed as a response to the criticism of the views of Raimond Sebond, the author of the theological treatise *Natural Theology* (around 1430), translated into French by Montaigne. The objections against Sebond, raised by some high-born ladies in France, were twofold; one set was aimed at Sebond's effort to reconcile faith and reason – Montaigne agrees with this criticism but defends Sebond's stance, proposing fideism. The second set of objections concerns apparent inconsistencies in Sebond's arguments that did not produce a satisfactory conviction. This was an opportunity for Montaigne to display a whole battery of sceptical ammunition with the aim of showing that a consistent argument is, in principle, impossible. "Let us try and see" he says, "whether a man has in his power any reasons stronger than those of Sebond – whether, indeed, it is in man to arrive at any certainty by argument and reflection" (Montaigne 1993, 12).

The verdict is obviously negative and Montaigne then examines the whole variety of sceptical doubts concerning the senses and rational judgment, plus the possibility of a rational means for establishing the existence of God (undermining thus the traditional scholastic arguments for God's existence). Montaigne advances a general definition of scepticism: "[the sceptics] use their reason for inquiry and debate but never make choices or decisions. If you can picture an endless confession of ignorance, or a power of judgment that never, never inclines to one side or the other, then you can conceive what Pyrrhonism is" (Montaigne 1993, 72). This definition reflects the "Sceptical Phrases" formulated by Sextus Empiricus – especially the phrases, "I determine nothing"; "Everything is undetermined"; and "Opposed to every account there is an equal account." Sextus eventually concludes, "I suspend judgment … the intellect is suspended so as neither to posit nor to reject anything because of the equipollence of the matters being investigated" (*PH* I, 196, 49).

Montaigne exposes the sceptical views concerning the unreliability of the senses and thus undermines the traditional Aristotelian conception in which senses provide the first information – information that is incomplete but not deceptive. Montaigne looks one by one at the ten modes of Aenesidemus; the ninth mode, which suggests the role of the frequent experiencing of certain phenomena in accustoming us to what is repeatedly observed, is especially interesting in connection with Hume. For instance, the sea is striking to someone who sees it for the first time but it does not excite us if it becomes a customary sight. One may consider this mode as the first timid precursor of Hume's idea of association by which repeated observation of regularities gives rise to our custom to expect the same connections in the future (and to our belief in its necessity). For Montaigne, specific defects in sense experience are finally overshadowed by a general doubt; we cannot know that our five senses are sufficient for the adequate picturing of objects. Maybe we lack some faculties that are necessary for perceiving nature and "the lack of such faculties entails our ignorance of their true essence"; therefore "the senses do not embrace an outside object but only their own impressions of it; therefore … the appearance [is] not a property of the object but only the impression and feeling of the senses" (Montaigne 1993, 173, 185–6). This sceptical motive anticipates the turn of early modern philosophy to phenomenalism, most strongly expressed by Descartes and Hume. For the Pyrrhonians, we are enclosed in the world of appearances, yet accepting this fact is no cause for despair, quite the contrary, for acceptance brings us peace of mind; Hume, of course, went further and introduced a positive alternative represented by instinctive beliefs that cannot be shaken by any kind of sceptical argument.

Montaigne further investigates whether reason can improve our situation and, unsurprisingly, reason also fails the test. The sceptical scrutiny of our rational dispositions unfolds from the five modes of Agrippa. Two of them (1 and 3) depict common tendencies of our use of argument; the mode deriving from dispute overlaps with the third mode from relativity; both show how even the strongest convictions may suddenly change, how many contradicting convictions exist among men and how powerless we are in deciding between them. Montaigne echoes these modes by saying: "How frequently we change our ideas! What I hold and believe today, I hold and believe with the totality of my belief … But – not once, not a hundred times, not a thousand times, but every day – have I not embraced something else with the same resources…?" (Montaigne 1993, 141–2). Given that there are many lunatics among us, reasoning does not have much authority; but even "the most learned, the best-endowed and cleverest of men never agree about anything, not even that the sky is above our heads", comments Montaigne with irony (Montaigne 1993, 142). Descartes, of course, offered a univocal criterion of how to judge propositions, namely the clarity and distinctness of ideas that we recognize due to the fact that we all have the same natural light of reason; Hume argues that our natural instincts, common to all, lead us safely in matters of life and in matters of cognition regardless of how reflection undermines them.

The fundamental sceptical assault on man's rational capacities, however, is contained in the remaining three modes, later shortened to two modes of reciprocity and infinite regress. Here, the classic criticism of induction and justification is displayed; no item of knowledge can be self-supportive but instead needs something else to support it: "…nothing is apprehended by means of itself … if that by means of which something is apprehended will itself always need to be apprehended by means of something else, they throw you into the reciprocal or infinite mode" *PH* I,

179, 44).[2] The result of this scepticism with regard to reason is unequivocal; we have to accept that "this malady is … past all cure. Reason always hobbles, limps and walks askew" (Montaigne 1993, 144). To sum up, all sceptical objections regarding the reliability of our senses and reason destroy the possibility of adequately understanding the world; Montaigne's verdict is that "the senses themselves being full of uncertainty cannot decide the issue of our dispute. It will have to be Reason, then. But no Reason can be established except by another Reason. We retreat into infinity" (Montaigne 1993, 185).

Montaigne then proceeds to another area of doubt where he considers the motive of a continuous dream which was later made so famous by Descartes. Montaigne, however, comes to different conclusions. He compares the difference between sleep and waking with regard to knowledge and argues that (a) the two states cannot be distinguished and life can be a continuous dream: "why should we therefore not doubt whether our thinking and acting are but another dream; our waking, some other species of sleep?" (Montaigne 1993, 180). Then he claims that (b) in sleep our cognitive faculties are somewhat darker but (c) the difference is not so great. Even if man were of the opinion that he makes rationally coherent propositions (as he cannot in principle) then the difference between the state of sleep when out mind is blurred and the waking state when man is "never so wide awake that it can cure and purge those raving lunacies …" (Montaigne 1993, 180) is negligible. Unlike Descartes, who uses the dream argument for establishing the indubitable certainty of *ego*, Montaigne uses it as an illustration of overwhelming doubt, unescapable whether we sleep or are awake: "the senses deceive our intellect; it deceives them in their turn" (Montaigne 1993, 179). Montaigne does not accept the softer, Academic version of scepticism and objects, "how can they [the Academics] bring themselves to yield to verisimilitude if they cannot recognize verity? How can they know there to be a resemblance to something the essence of which they do not know? We judge entirely, or entirely not" (Montaigne 1993, 141).[3]

Montaigne goes along with the sceptical conclusion that we are acquainted only with appearances and claims that "the senses do not embrace an outside object but only their own impressions … Those impressions and that object are different things" (Montaigne 1993, 185–6). Likewise, reason is an epistemologically constructive – but subjectively relative – power: "by reason" he says, "I always mean that appearance of rationality which each of us constructs for himself … [it] is like a tool of malleable lead or wax: it can be stretched, bent or adapted to any size or to any bias" (Montaigne 1993, 144). Montaigne set the scene for centuries-long debates in which philosophers tried to find some solution to the crisis. A remedy was sought in all possible venues of philosophy – in scholastics, in fideism and the Augustinian tradition, in mitigated scepticism, by exposing scepticism in its most radical form but finding the new certainty in the thinking self or, as in Hume's case,

[2] Thorough studies on Pyrrhonism and the sceptical modes can be found, for example, in Annas and Barnes (1985) and Hankinson (1998).

[3] Contrary to Montaigne's own words, Bermúdez Vázques (2015) claims that Montaigne was in fact closer to Academic scepticism and not to Sextus.

in accepting the Pyrrhonian verdict but finding the solution in naturalism. All these attempts to solve the Pyrrhonian crisis created a unique intellectual atmosphere which was full of new and revolutionary ideas that helped to form the new paradigm. Montaigne triggered off an amazing development in philosophy.

3.4 Is There a Remedy for Scepticism?

Could anything provide a consistent remedy for these lethal sceptical conclusions? Scepticism had shown that man's situation was disconsolate – "is it possible to imagine anything more laughable than that this pitiful, wretched creature – who is not even master of himself, but exposed to shocks on every side – should call himself Master and Emperor of a universe …!" (Montaigne 1993, 13). While Descartes celebrates this privileged position of man, Montaigne considers it a shameful vanity. He insists on the necessity of God's guidance and, inspired by St. Augustine, appeals to Divine illumination. If we accept the gift of faith then God by his grace grants to us the possibility of understanding the world, of grasping its structure and beauty: "our human reasoning and concepts are like matter, heavy and barren: God's grace is their form, giving them shape and worth" (Montaigne 1993, 11). Faith, however, is not an aid for the improvement of reason for cognitive purposes; Pyrrhonism showed man "naked, empty, aware of his natural weakness" (Montaigne 1993, 74). Montaigne, like Aristotle before and Locke after, uses the example of a (wax) writing tablet, yet in a different, fideist sense: "he [Man] is a blank writing tablet, made ready for the finger of God to carve such letters on him as he pleases" (Montaigne 1993, 74). Hume picks up the theme of the despair felt when one encounters the wretched state of human reasoning faculties, and discusses it without any recourse to God.

In the *Apology*, Montaigne elaborated a fideist conception of faith. He saw in fideism the way to save religion in a situation in which the validity of the traditional proofs of God's existence were being discredited by sceptical attacks on reason. Montaigne recommended transferring faith to the realm of the heart – "only faith can embrace, with a lively certainty, the high mysteries of our religion" (Montaigne 1993, 3) – and to elevate the status of faith in supernatural revelation. Similar phrases appear in Pierre Bayle's *Dictionary* and in Hume's *Dialogues* on behalf of Demea. On the fideist position, the certainty of God's existence is thus beyond the reach of both reason and scepticism. Montaigne's endorsement of fideism also played an important role in his defence of Catholicism against the 'poison' of Calvinism that undermined traditional Church authority; the reformed religion was based on the individual's contract with a God who demanded individual accountability for deeds and sins. Against this, Montaigne promoted the Pyrrhonian respect for customs and traditions. Man who finds himself amidst upheavals which shake the hitherto accepted order is disoriented, uprooted: "as I do not have the capacity for making choices myself" – says Montaigne – "I remain where God put me. Otherwise I would not know how to save myself from endlessly rolling. And thus, by God's grace, without worry or a troubled conscience, I have kept myself whole, within the ancient beliefs of our religion, amidst all the sects and schisms that our century produced" (Montaigne 1993, 149).[4] The Pyrrhonian ideal of peace of mind and inner calm is hereby achieved. This tradition, based on the preference of faith over reason, was further developed by Montaigne's disciple Pierre Charron and continued via Blaise Pascal to the sceptics Pierre Gassendi, Simon Foucher and Pierre Daniel Huet, on to Pierre Bayle and eventually to Kierkegaard.

However, let us return to the problem of knowledge. *Scepticism was a double-edged sword that had both a liberating and a paralyzing effect on epistemology.* On the one hand, it helped to free man from the traditional authority of scholastic doctrine; scepticism thus *boosted his self-confidence*, resulting in the release of his intellectual potential. On the other hand, it *undermined man's self-confidence* by illustrating the unreliability of his cognitive faculties, his senses and his rational judgment alike. Metaphorically, by adopting scepticism man cut off the branch he was sitting on; scepticism showed the impotence of man's own mind. That clashed with the newly-acquired confidence and hunger for knowledge. This clash led to the *crise pyrrhonienne* that showed man as an intellectually pitiful creature. The ancient Pyrrhonian solution consisting in resignation, in accepting these cognitive limitations and in detachment from the world as a way to finding peace of mind, threatened to destroy this grand epistemological project.

Yet what kinds of solutions to the Pyrrhonian crisis were available? The whole intellectual situation was very unstable, with the traditional scholastics still enforcing its influence, with the rationalist attempts to defeat scepticism, often linked to various metaphysical conceptions that were not quite acceptable to the traditional scholastics, with the emerging sceptical empiricism linked to a stronger or milder fideism. It was a mixture of influences where various groups not only attacked each other but also suffered many internecine tensions. In all this turmoil, Descartes and Hume held the most radical though opposing positions, the former opening the door to modernity – to the autonomy of the individual mind – and rejecting scepticism and the latter – already following the modern path – completing the Pyrrhonian crisis. They both understood the seriousness of the sceptical challenge and realized its destructive impact on epistemology.

3.5 The Rationalist Solution

Descartes was a central figure in all the debates – philosophical, theological and scientific. His solution to the sceptical problem did not earn much admiration from his contemporaries on either side, impressing neither the rationalists

[4] Sects for Montaigne are various forms of Calvinism; but for Descartes 'a sect' meant the sceptics.

(neo-scholastics) or the empiricists. The rationalist anti-sceptical alliance[5] drew on various sources: on the Aristotelian scholastics (Caterus and Bourdin, for example); on the Augustinian tradition (Malebranche); and it absorbed further influences, such as Mersenne, who supported Descartes' rationalism but was also open to the newly rising experimental "sceptical" science of Gassendi. The criticism of Descartes rendered by this diverse group centred on several issues, amongst the most important being Descartes' rejection of the deeply-ingrained belief in the unity of the soul and the body, his intellectualist conception of the mind and his subjectivist grounding of knowledge in *ego cogito*. From the viewpoint of science, Descartes' contribution was acknowledged with regard to his rejection of the Aristotelian doctrine of cognition as a teleological process, and of the theory of the four elements of which everything was made. However, the critical voices were stronger that the appreciative.

For Aristotle the human soul, *anima*, was the organic principle of life which permeated (or *animated*) our nature, including our intellect and our body; and the unity made the person whole. Descartes' conception of the mind, *mens*, as a purely intellectual substance appeared to the traditionalists as an untenable, artificial speculation. Mersenne, for instance, asks Descartes: "you so stoutly resisted the claim of all bodies to be [more than] phantasms in order that you might be able to draw the conclusion that you were merely a thinking being … [but] what if that were a body which by its various motions and encounters produces that which we call thought?" (Mersenne 1934, 24–5). From another angle, Hobbes raises a similar objection: "all Philosophers distinguish a subject [*subiectum*] from its properties and activities, i.e. from its properties and essences. Hence it is possible, for a thing that thinks to be a subject of the mind, reason or understanding and hence to be something corporeal" (Hobbes 1934, 61). For Bourdin, Descartes' identification of the self with thinking only shows "the perception of the existence of mind, and fails to reveal its nature" (Bourdin 1934, 149). The "nature" of mind simply could not, for Descartes' contemporaries, be just *thought* – it was too alien to the discourse of the time, scholastic or sceptical.

Another important criticism of Descartes unfolds from his declaration that the individual mind, the *ego cogito*, is the only certainty that can resist the sceptical assault. Again, in his time this was seen as either a sign of vanity or utter desperation; how can a single human mind be the only indubitable certainty, standing out of the context of the whole of Being, isolated from the certainty of God's creation that unites the heavens, nature and our souls? How can the individual mind/intellect be the foundation of epistemology? It was seen as ridiculous. These objections to Descartes are thus understandable from the critics' own traditional beliefs, preventing them from an open-minded consideration of the novelty of Descartes' philosophy. Descartes was, paradoxically, blamed for *excessive scepticism* for ending in the *Ego* alone, and for *atheism* as a result. The *cogito* as a new beginning of philosophy

[5] I want to mention once again that I do not include in this list those philosophers who were not involved in the sceptical controversy, above all Leibniz and Spinoza; despite disagreements on many specific issues they considered satisfactory Descartes' founding of knowledge in metaphysics, thus avoiding the danger of scepticism. For a detailed analysis, see R.S. Woolhouse 1993.

and a revolutionary step to modernity was not only not appreciated; it was even not comprehended. Malebranche, for instance, a supporter of Descartes' rationalism, when he saw how badly it was received, suggested as a remedy to replace the *Ego* with God as the foundation of knowledge, thus destroying the core of Descartes' achievement.

Descartes was especially unhappy about the animosity of Bourdin, the leading Paris Jesuit, who, as he knew, would influence the official attitude of the Order towards his philosophy. Descartes protested against Bourdin's charge of excessive scepticism and emphasized that his intention was to eliminate scepticism and not to deepen it, as Bourdin claimed. *Descartes considered Pyrrhonian scepticism a perversity* endangering his whole project to establish the rational foundation of universal science. In his *Reply* to Mersenne's *Objections* he wrote, "although I felt disgust by serving up again this stale dish [old cabbage], I could not for the above reasons [the threat of scepticism] refuse to allot to this subject one whole meditation" (Descartes 1934a, 31). Descartes realized the disastrous impact of scepticism on epistemology and was determined to overcome it by establishing an indubitable certainty that is immune to sceptical attacks; by logical arguments, eliminating one alleged certainty after another, he ended with the self-justifying *ego cogito*; but this was not an option for either the traditional scholastics or the sceptical empiricists.

The inconsistency between the first parts of Descartes' *Meditations* and the parts in which he withdrew from the position of *ego cogito* – and in which he denounced almost all that he had proclaimed before – did not change the severity of these criticisms. It may seem strange since Descartes, especially in the sixth *Meditation*, claims that "nature also teaches me … that I am not present in my body merely as a pilot is present in a ship; I am most tightly bound to it, and as it were mixed up with it, so that I and it form a unit" (Descartes 1979, 117).[6] He thus denounces the dualism of mind and body and the epistemological autonomy of the intellect. He also makes room in epistemology for the power of our nature – senses and imagination – and recognizes some epistemic contribution from our natural instincts. Finally, Descartes puts the certainty of the self and the possibility of true knowledge into the hands of God. The rehabilitation of the wholeness of a person and the wholeness of being is guaranteed by God's perfection and omnipotence; Descartes says that his nature complies with "the complex of all that God has given me" and that everything belongs to "the order of created things established by God" (Descartes 1979, 116).

However, the scholastics were not impressed by Descartes' rehabilitation of God. As discussed in 2.6., they recognized that God was "awarded" this role only after Descartes was ensnared in the trap of the autonomous cognizing subject.

[6] The metaphor of a pilot in a ship is mentioned by Aristotle in *De anima* (*Peri psyche*). Aristotle analyzes the unity of the soul and the body, the soul being the principle of life. Descartes in this *Meditation* expresses the same idea as Aristotle but he objects to the metaphor of a pilot and a ship as being too loose, not capturing the strength of the bond between them. Later Kant uses the same metaphor in different circumstances – in his criticism of Hume's psychologism that eliminates the role of a pilot – reason – from knowledge.

Despite the explicit phrases about the foundational role of God, *Descartes' God has a purely epistemological purpose*. God made the world such that it can be transparent to our reason: in other words, it can be the subject-matter of mathematics. In his ontological proof of God, Descartes begins with a comparison of the certainty of the idea of God and the ideas of some mathematical entities (figures or numbers); the goal of this comparison is to prove that "I ought to hold the existence of God with at least (*sic*) the same degree of certainty as I have so far held mathematical truths" (Descartes 1979, 103); for instance the certainty that the sum of the angles of a triangle is equal to the sum of two right angles. Placing these two proofs on the same level makes the existence of God just another proposition in demonstrative reasoning.

Descartes then comes to the crucial difficulty implied by his previous radical arguments, i.e. the impossibility of bridging the gap between the mind and the world, and between certainty and truth; "for my thought (*cogitatio*) imposes no necessity on the existence of things" (Descartes 1979, 104). For this reason, he proceeds to the ontological proof: "from my inability to think of God as non-existent, it follows that he really does exist" (Descartes 1979, 104). It is the only idea the existence of which is not dependent on my way of thinking, but on the contrary, my thinking and all knowledge depend on His existence; the idea of God is supreme whereas the certainty of the angles of a triangle may be doubted at moments when we are overwhelmed by pessimistic thoughts concerning our thinking abilities. But most importantly, Descartes can now insist that truth is identical with being. The bridge from subjective certainty to truth is built: "from this [the existence of God] I have gathered that whatever I clearly and distinctly perceive is necessarily true" (Descartes 1979, 107). Despite the fact that Descartes returns to metaphysics, I think that deep down he never gave up the belief that it is not the divine light but the *light of our reason that 'illuminates' (rationally comprehends) the world*. True, Descartes condemned metaphysics only to come back to it later, but his embracing it was not sincere, as the scholastics saw. The main reason for Descartes' turn to metaphysics appears to be *the impossibility of getting from the self to the external world, from subjective certainty to truth*, not his religious belief.

3.6 The New Pyrrhonians

The rejection of Descartes' rationalism played a major role in the broad and diverse alliance of the new Pyrrhonians, a group that formed around Pierre Gassendi and Samuel Sorbière and continued in the next generation to Pierre-Daniel Huet and Simon Foucher. Gassendi and Huet were two of the most prominent proponents of the opposite solution to the Pyrrhonian crisis: *they accepted the fact that Pyrrhonism could not be defeated* and tried to find some operational space for science within its framework. Their approach was based on empiricism, an approach that advocated experimental science and called for modesty in our claims to knowledge. Obviously, this position takes us nearer to Hume.

Their criticism of Descartes was not univocal – almost nothing was in this unsettled time. Most of the new sceptics and scientists appreciated Descartes' anti-Aristotelian position in science, just as the rationalists did. Descartes' mathematical science (or any other modern form of science) could not function within the Aristotelian organic conception of nature and the qualitative hierarchy of being. Descartes rejection of the theory of substantial forms and final causes in physics (deemed occult by him) was also appreciated. Descartes confessed to Mersenne, in confidence, that his *Meditations* contained all the foundations of the new physics, and expressed the hope that readers would gradually get used to those principles and recognize their truth, *before they noticed that they destroyed the principles of Aristotle*. Thus Descartes' plan was from the start to reject both scepticism and metaphysics.

At the same time, though, the sceptics rejected Descartes' intellectualism, the certainty of the thinking self and the conception of innate ideas; in this respect, their criticism overlapped with that of the scholastics. The overlap is possibly strongest in the sceptics' rejection of Descartes' definition of mind in terms of rational thought. It was an ironic situation: although the sceptics strove to reject Aristotelian natural philosophy they still clung to the Aristotelian account of the soul whereby the soul animates the living body – it is the *reality* of body. They also perceived Descartes' conception of mind, as formulated in the second *Meditation*, as artificial. In his *Objections* Gassendi joins the other critics and expresses his disillusionment with Descartes: "I believed that I was addressing the human soul, or that internal principle by which a man lives, feels, moves from place to place, and after all I was only speaking to a mind, which has divested itself not only of a body, but of the soul itself" (Gassendi 1934, 141). Thinking is, for Gassendi and other philosophers of that time, only one of the attributes of mind and does not capture its *nature*; repeating the objection voiced by Bourdin, the nature of the soul was for Gassendi a unity of intellect and *sensibility* that belongs to the body. If we think of wax, Gassendi argues, and consider its features that we see, touch, etc., we conceive of something that we cannot do without eyes or hands.

Gassendi could not accept Descartes' argument that the vital attributes of wax are its extension in space, acquired by intellectual insight. Descartes accuses Gassendi of employing rhetorical wiles – or humbug – against him that is based on a crucial misunderstanding of his conception of mind; "[the fact that] the entire testimony of the senses must be considered to be uncertain, nay, even false … is necessary for the comprehension of my meditations, that he who cannot admit that … is unfit to urge any objection to them that merits a reply" (Descartes 1934b, 207). Gassendi not only disagrees with Descartes' devaluation of the senses in acquiring knowledge; he does not understand the definition of *ego* through the self-reflexive act (*cogito*); as Descartes says, I may see what I think is wax – it may or may not be wax – but "it is not possible when I think I see [say, wax] that my conscious self should not be something" (Descartes 1979, 74). This misunderstanding is manifest in Gassendi's claim that *ego cogito* could be easily replaced by *ego ambulo* (I walk); and a similar remark can be found in Hobbes' *Objection*, where he suggests using the phrase "I experience" instead of "I think". For both, thinking was inseparable from perceiving and from other modes of experience or movements that assumed the existence of body. However, for Descartes, perceiving, imagining, willing or walking are primarily *states or activities of which we are conscious*.

Gassendi and Hobbes may be right that walking or experiencing are connected to the corporeal substance; but walking or experiencing may also be only a dream. It is the awareness of these acts that points out to the thinking self. Popkin mentions an even more blatant misinterpretation of Descartes by Huet, who argued from a strongly hostile position that *ego cogito* presupposes a certain time sequence and therefore involves memory – another unreliable human faculty. Huet (in his *Censura philosophiae Cartesianae*, 1689) concluded that Descartes' *cogito* amounts to nothing more than "I may have thought, therefore perhaps I may be" (Popkin 2003, 209). The main mistake of the sceptical empiricists consisted in the assertion that Descartes' *ego* had to have corporeal attributes; but once mind is mixed with bodily sensations it was easy for them to claim that Descartes' *ego* loses the privilege of being an unshakable certainty. As soon as the sceptics included sensory perceptions into the nature of the mind the certainty of *ego* could not be self-justifying. Thus after this unfair manoeuvre against Descartes, the sceptics could solemnly declare him a dogmatist. Despite these negative reactions Descartes was a focal figure among scholars; everybody read him and the disputes about his philosophy shaped further philosophical developments.

The new Pyrrhonians adhered to the principle that the soul does not exist without the body and that *we cannot achieve knowledge without using first the senses*. They took seriously the Pyrrhonian display of the weakness of our mind and hoped to solve this problem by making a few concessions. Following Pyrrhonism, they (more or less) limited knowledge to appearances, replaced certainty with probability, advocated caution in generalizations and promoted modesty as a general attitude. This approach may seem very close to the ancient sceptics but there was one huge difference that marks the different epochs in which these two kinds of Pyrrhonians lived. *The primary concern of the new Pyrrhonism was science* (as it was for Descartes) free of any metaphysical assumption and charged with optimism; a whole new world was opening in scientific knowledge based on unprejudiced careful observations, experiments and mathematics, conducted by the individual mind. The goal of the old Pyrrhonians was aimed inwards; sceptical inquiries led to the suspension of judgments, withdrawal from life and inner peace of mind. Achieving this state of mind, linked to a tolerant and humble attitude to life, was the ultimate goal.

But did this group of modern sceptics succeed in reconciling scepticism and science? Did they blunt the blade of the destructive power of Pyrrhonism so that science could be rationally legitimized? Gassendi formulates the methodological rule that knowledge starts with experience and claims that direct observation cannot deceive us as long as we limit knowledge to how things appear to us. This claim is fully consistent with Pyrrhonism. However, science requires more than this – scientific propositions involve judgments that reach beyond directly apprehended empirical data. In this phase our reasoning is susceptible to error: "At this point" says Gassendi, "I have no desire to begin a controversy about the trustworthiness of the senses; for, if there is a disposition of falsity, it is not in sense, which is merely passive and has to do only with things that appear and must appear in the way they do and owing to their own appropriate causes; it resides in the judgment or in the mind…" (Gassendi 1934, 192–3). If we cautiously record the observational data – how things appear to us in terms of colour, taste, touch, size etc.[7] – we could not yet make perceptual judgments that are necessary for knowledge. How can a Pyrrhonian defend any kind of judgment?

Ralph Walker (1983) shows that Gassendi has to allow for some justification of such judgments and that he does so on the basis of their accordance with the majority of mutually consistent perceptual judgments. A true Pyrrhonian would reject this strategy as dogmatic and appeal to the principles of non-assertion and indeterminacy; opposed to every account there is an equal account and we can only say of an assertion that now, when we utter it, we feel in this way with regard to these matters under investigation; in Sextus' words, "I now feel in such a way as neither to posit dogmatically nor to reject any of the things falling under this investigation" (*PH* I, 197,49). Gassendi himself, at the end of his *Objections* to Descartes, demonstrates this sceptical manner and understates the importance of his arguments; he says, "For as, when some food is pleasant to my palate, I do not defend my taste, which I see is offensive to others, as being more perfect than anyone else's; so when my mind welcomes an opinion, which does not please others, I am far from holding that I have hit upon a truer theory … each enjoys his own opinion" (Gassendi 1934, 203). In practice, though, Gassendi considers the consent of the informed majority satisfactory since we all receive the same sense information; for him, "the discovery of occasional illusions cannot cast doubt on the reliability of perceptions in general … illusions are only occasional, or only in abnormal circumstances" (Walker 1983, 324). This may be a slightly inconsistent way out of Pyrrhonism but there are more pressing problems looming for the sceptical empiricists.

Can the claim that knowledge is the observation of *how things now appear to us* be acceptable to science? Is science not always concerned with describing the real world, real objects and physical laws, real planets and their motions, the forces that make up the world and the universe? Are generalizations not constitutive for scientific theories? Gassendi would have to explain how we can relate our knowledge to anything beyond a subjective report on appearances. He would have to establish what Descartes (before his metaphysical turn) failed to establish; firstly that my perception of "redness" is a real attribute of a thing (a substance) and secondly (and crucially) that there are independently existing things that affect our senses. Gassendi, as we could see in his polemics with Descartes, did not accept the implications of Descartes' *epoché* that puts into parentheses everything except the awareness of me thinking. He accepts the ancient Pyrrhonian position in which we limit the scope of our knowledge only to how things appear to us and deny the possibility of revealing the truth about them, but *we do not doubt their existence*. In the ancient

[7] Gassendi considers some qualities – magnitude, size, shape – more constant than others like taste, touch, hot cold; in other words, sensible qualities. Yet ultimately all qualities are compounds or combinations of the elementary particles, the atoms.

philosophical framework such radical ontological doubt was not yet possible; however, it should have been on Gassendi's agenda.

He avoids direct confrontation with phenomenalism and inclines rather to a pragmatic realist position, especially when predictions in science come into play; in this context, Gassendi thinks that science is concerned with the *real intrinsic character of things*. However, this takes him, as Walker argues, straight into the sceptical trap that Descartes faced and in which Gassendi was not even aware of being caught – a trap from which we can escape with the help of metaphysics or dogmatism, both strictly rejected by Gassendi. In contrast with this interpretation, Popkin thinks that Gassendi is consistent and says, "Gassendi advocated total scepticism about the world beyond appearances" and in science "we describe these scientific objects (the atoms) in terms of the qualities found in experience" (Popkin 2003, 122–123). There is certainly a tension between Gassendi's commitment to appearances and his work as a scientist who studies the real world and its general laws, a tension that is open to discussion. Generally, the philosophical position of the Gassendian circle and its followers – most importantly François de La Mothe Le Vayer, Samuel Sorbière, Simon Foucher and Pierre-Daniel Huet – can be described as mitigated epistemological scepticism.

3.7 Mitigated Scepticism

Mitigated or constructive scepticism adopts some Pyrrhonian themes; the weakness of our reason and the resultant caution in judgments; the limitation of our judgments to appearances thus relinquishing claims to knowledge of the hidden natures of things; humility in confrontation with other persons' opinions instead of pride and vanity; the rejection of any outside authority – here represented by scholastics – on the grounds of dogmatism. All these characteristics fit perfectly with Hume's description of scepticism. The experimental method that Gassendi proposed, rid of metaphysical issues and admitting no 'hypotheses', may have influenced Newton and the further development of experimental science. However, if we *limit* scepticism to these items we arrive at the ideal of mitigated or true scepticism, which, for Hume, is not capable of exposing the epistemological contradictions; as argued in the next chapter, Hume's mitigated scepticism is not applicable to epistemology but is intended for moral science and "common life". Mitigated sceptics of the Gassendian provenance, by applying mitigated scepticism to epistemology, silently assumed that science is not about appearances but reality, though they granted a high degree of fallibility to our judgments. Hume understood that constructive scepticism had very little to do with Pyrrhonism except the recommendation to modesty and criticism. Hume read the then very popular Huet's *Traité philosophique de la foiblesse de l'esprit humain* (1723), translated into English by 1725. Huet adopted a strong sceptical tone but never went as far as Hume – he held the probabilistic attitude to knowledge combined with fideism: in his *Treatise* he aims at "showing that Truth cannot be perfectly and with certainty known to human Understanding"

(Huet 1725, 9). Hume, by contrast, faced the consequences for knowledge of Pyrrhonism and admitted there was no cure for scepticism except switching to a different kind of discourse – from rational reflection to natural instincts.

Gassendi and Huet's mitigation of Pyrrhonism implies that that we keep a blind eye to the devastating effects of Pyrrhonism on knowledge. On the one hand we proclaim that knowledge cannot reach anywhere beyond appearances; on the other hand we assume that science, though probable and fallible, is about nature and the universe. The latter downplays the sceptical aspect in order to strengthen science and does so without qualms. As mitigated sceptics declare this strategy to be *functional* and its result *satisfactory*, we do not need more for life or science. In one passage Locke refers to Descartes' phenomenalism and calls him a dreamer who affirms that "all we see and hear, feel and taste, think and do… is just a dream" and he continues, "if he pleases, he may dream that I make him this answer, That the certainty of things existing in *rerum natura* when we have the testimony of our senses for it is not only as great as our frame can attain to, but as our condition needs… Such an assurance of the existence of things without us is sufficient" (Locke 2011, 634). To mock Descartes' intellectualism he recommends to the dreamer to put his hand to a furnace and then, "he may perhaps be wakened into a certainty greater than he could wish, that it is something more than bare Imagination" (Locke 2011, 635). Constructive scepticism defends a pragmatic attitude to knowledge, based on cautious empirical observations of perceptual data and remaining down to Earth when theorizing. Gassendi thus initiated a move to a sober experimental science that impressed Locke, Newton (who read Gassendi) and other empiricists.

French philosophers appeared to exert a strong influence on British philosophy and science at that time; Descartes' role had been crucial; Gassendi's mitigated scepticism influenced generations of sceptics; in science, Newton was influenced by Gassendi's theory of light as composed of extremely small atoms of an exceptionally rarefied nature. Boyle, for instance, appreciated Descartes' anti-Aristotelian line but disputed the Cartesian theory of matter, his laws of motion and the possibility of their quantification; he was inspired by Gassendi's atomism according to which everything ("Catholick matter") could be reduced to a single element (*corpuscule*) that gave rise to all diverse clusters and chemical qualities of matter (according to the relation between the pressure and volume of gas in the air). Boyle rejected the theory of the four elements (Aristotle) and also Fludd's alchemical concept of three elements (sulphur, salt and mercury) according to which they – when combined in various balances – create all there is. At the same time Boyle adopted the experimental method in science and carried out many experiments in his laboratory with the occasional assistance of Locke. Boyle summarized his discoveries in his famous *Sceptical Chymist* (1661), considered the foundation of modern chemistry as an independent discipline, no longer subservient to medicine.

On a more philosophical level, the importance of Glanvill for Hume should be noted. Glanvill formed his thought under the influence of Descartes (although he rejected Descartes' metaphysical theory) but later preferred the French sceptics; he formulated his principles, typically combining strong anti-Aristotelianism (for which he was charged with atheism), mild fideism and empiricism, in his *Scepsis Scientifica* (1665), an expansion of his earlier work, *The Vanity of Dogmatizing*. Glanvill's scepticism anticipated Hume's through his polemics against the empirical justification of causality and necessary connection; according to Glanvill, they cannot be considered demonstrative knowledge. With respect to naturalism, Hume could have found some fragments of inspiration in other philosophers' writings. Popkin suggests that the idea of naturalism as opposed to scepticism may have come from Jean Pierre Crousaz' *Examen du Pyrrhonisme ancien & moderne* (1733) which sums up his other earlier writings on scepticism. Crousaz considered scepticism a menace, a contagion spreading throughout Europe. He illustrated the madness of scepticism with the fact that sceptical doubts contradict our natural feelings; scepticism is irrefutable but unbelievable. Crousaz thought that this opposition between scepticism and feeling discredited Pyrrhonism but Hume turned the argument around and argued that such opposition was a perfect description of the human condition when both parties must be given their own rights, despite being directly incompatible. This was the first step that Hume took towards the correction of Pyrrhonism.

Locke especially was influenced by Gassendi – he was in contact with the circle of Gassendi's followers during his stay in France and probably met Huet (though his trip was officially to enhance his medical education). Popkin demonstrates a similarity between Locke's and Gassendi's texts and says it is especially striking when Locke's *Essay* is considered in terms of Gassendi's *Syntagma Philosophicum* (1658, posthumously); Popkin even suggests calling Locke a latter-day Gassendist[8]; The principles of experimental science, free of metaphysics and based on observation, lacking the ambition to reveal the utmost secrets of nature, formed a perfect philosophical and methodological basis for the new science that had its institutional centre in the Royal Society, gathering the best scientists (and philosophers) of the time. The intellectual source of this new stance, which paved the way to modern science and philosophy, was in France, but in due course experimental science gained greater influence in Britain.

3.8 Metaphysical Assumptions

It would be tempting to end the chapter here. The revival of Pyrrhonism contributed to the disintegration of Aristotelianism both in philosophy and in science at a time that was ripe for such a change. However, on closer examination another aspect of this development is worthy of consideration. As stated in this chapter, scientists and philosophers of this period explicitly highlighted one common enemy – metaphysics of any form. Yet none of them adopted the main philosophical implication of the

[8] Michael & Michael (1990) did excellent work in tracing the ways in which various philosophical themes of Gassendi and his followers influenced Locke. For further aspects of the influence of Gassendi see Lennon (1993).

disintegration of all metaphysical certainties – the autonomous status of the individual mind and the resulting loss of external objectivity. Obviously, phenomenalism was endangering science. Yet was the concern for science so dominant that it would overshadow philosophical considerations? There may be another explanation for the lack of philosophical consistency among the empiricists. Under the sober experimental surface of the emerging natural science they still relied on *silent metaphysical assumptions*.

In the previous chapter Newton's insistence on the experimental nature of science was addressed. But the metaphysical framework of Newton's science is another problem to be discussed. Although theology and alchemy were supposed to be kept on different shelves from science and not to interfere with scientific explanations, they formed an undeclared yet self-evident grounding of the observed phenomena and processes. Until recently, the extent of Newton's writings on alchemy and theology was not well known – and it could not be because these writing were dispersed around the world; in 1936 they were divided into three hundred and thirty lots and sold at an auction at Sotheby's. Prior to that they were in the possession of the family of Newton's niece, and offered to the University of Cambridge and to the British Library; both institutions declined to take them on the grounds that Newton's reputation as a scientist would be stained. It is thanks to the Newton Project that these manuscripts are being traced, scanned and categorized. This organization is dedicated to publishing in full an online edition of all Newton's writings – the count is so far over six million words. It has become apparent that Newton's texts on alchemy and theology many times exceed his scientific writings. But quantity is not the main measure of the influence of theology and alchemy on Newton's science, and we still have to wait for a proper evaluation.

Tessa Morrison in her excellent book (2011) describes the long way of these manuscripts to public recognition, and argues for the underlying unity of all Newton's writings. Historically, John Maynard Keynes (one of the purchasers at Sotheby's) was among the first to appreciate their importance and he called Newton the last magician "because he looked on the whole universe and all that is in it as a riddle, as a secret which could be read by applying pure thought to certain evidence, certain mystic clues which God had laid about the world to allow a sort of philosopher's treasure hunt to the esoteric brotherhood" (Keynes 1972, 366). He believed that these clues were to be found partly in the evidence of the heavens and in the constitution of elements (and that is what gives the false suggestion of his being an experimental natural philosopher), but also partly in certain papers and traditions handed down by the brethren in an unbroken chain back to the original cryptic revelation in Babylonia. Newton, as Keynes argues, regarded the universe as a cryptogram set by the Almighty. Morrison supports this view and says that "Newton's deeply held religious convictions led him to search for the mystic clues which he believed that God laid about the world" (Morrison 2011, 5). She focuses on Newton's attempts to reconstruct the Temple of Solomon from biblical sources, arguing that for Newton *the universal order was laid out in the architecture of the Temple*; the Temple was a microcosm mirroring the universe. Theology, alchemy and science were all simultaneously studied and written on by Newton, "contrary to popular opinion, that

Newton's theological work, including chronology and prophetic interpretation was the work of an elderly and senile Newton" (Morrison 2011, 103).

Newton included his theological doctrine in his scientific writings; in the *General Scholium* (added to the second edition of the *Principia*) many metaphysical passages can be found that are of great importance. The following is among the strongest: "this [divine] Being governs all things, not as the soul of the world, but as Lord over all: And on account of his dominion he is wont to be called *Lord God Pantokrator*, or *Universal Ruler*" (Newton 1934, 544).[9] Similarly, Newton claimed that "all the diversity of natural things … could arise from nothing but the ideas and will of a Being necessarily existing" (Newton 1934, 546). He takes the universe and this world to be not only created by God but maintained by him (*Deus conservat mundus*); God continually re-establishes the universal order and harmony; he keeps the planets moving and rotating, thus preventing them from falling. In the Leibniz-Clarke correspondence of 1715–1716, Leibniz mocks this view, saying: "Sir *Isaac Newton*, and his Followers, have also a very odd Opinion concerning the Work of God. According to their Doctrine, God Almighty wants to *wind up* his Watch from Time to Time: Otherwise it would cease to move. He had not, it seems, sufficient Foresight to make it a perpetual Motion. Nay, the Machine of God's making, is so imperfect, according to these Gentlemen; that he is obliged to *clean* it now and then by an extraordinary Concourse, and even to *mend* it, as a Clockmaker mends his Work" (Leibniz 1717).

It is obvious that a person like Newton could have never doubted the independent existence of nature; the question rather is whether (and how) these two spheres – metaphysical and scientific – interacted. Stephen Snobelen, for example, claims that there was a close contact both ways; "an interpenetration existed at a fundamental level between the cognitive content of the theological and natural philosophical features of Newton's grand study" (Snobelen 2001, 22–3; see also Westfall 1982). According to him the theological and alchemical writings were intended as the esoteric knowledge – and the only knowledge that delivers *the truth*; while the experimental science was intended for the public and kept its focus within the empirically given, restricting itself to phenomena. Newton could be seen as reviving the old ontological concept of truth as given to us by God. Therefore, as Dobbs suggests, "reason and revelation were not [for Newton] in conflict but were complementary. God's attributes were recorded in the written Word but were also directly reflected in the nature of Nature" (Dobbs 2002, 6). By the way, Boyle, too, adhered to the view that there are supernatural cosmical qualities that transcend the mechanistic laws, and took that as a vindication of God's intervention in the world.

This applies even more to alchemy, which was supposed to be kept secret. Alchemy might have been the initiation into the mysteries of nature and the universe. The boom of alchemy in the Renaissance continued to the early modern era and alchemy found many supporters among natural philosophers; Gassendi was an alchemist and so too were Boyle and even Locke (to a lesser extent) and, of course,

[9] Newton – who was an anti-Trinitarian – draws in this quote on Deuteronomy 10:17: For the LORD your God is God of gods, and Lord of lords.

Newton. Newton's theory of aether was hailed by alchemists as an explanation of a non-mechanical action at distance. Newton formulated his aether theory around 1672 and submitted it in a memorandum to the Royal Society in 1675; there he explained gravity in terms of aether, responsible for holding planets in their orbits. He suggested that the whole universe might be filled with a weightless invisible elastic medium, comprising tiny undetectable particles, which is capable of propagating vibrations. This aether pervades the pores of all material bodies and is the cause of their cohesion; its density varies from one body to another, being greatest in the free interplanetary spaces. Newton described the principles of the cosmic aether in a unique letter of 1679 to Boyle (Appendix 1).

Atomism, too, did not serve just as a foundation of the new physics but was also used in alchemy. Conceived as basic micro particles that have a tendency to divide and ultimately to crumble to dust, atoms could endlessly be transformed into various chemical substances, according to their combinations and interactions. Boyle cannot be seen simply as the father of modern chemistry; his *Sceptical Chymist* outlines alchemical principles and procedures.[10] Minute particles of the universal matter give rise to the diverse substances of the world and changing these characteristics using chemistry could transform any material into any other one. Boyle carried out experiments in which he tried to transform base metals into gold and strove to prepare the 'philosopher's stone', the secret substance that would enable this. We know that Newton urged Boyle to refrain from publicly discussing his alchemical experiments. Boyle, before his death, sent to Locke what he believed was the procedure of the transmutation of metals into gold and Newton warned Locke against making any experiments unsupervised, since opening such secrets might be dangerous.

Alchemy was taken as the clue to the most profound truths, understood only by a few; groups and secret societies were formed guarding the secrets (experiments in transmuting metal into gold being forbidden by law until 1689, when Boyle successfully petitioned Parliament to lift the ban). Newton succeeded Boyle as the Grand Master of the Priory of Sion (following, among others, Leonardo da Vinci and Robert Fludd), a secret order sometimes said to be linked to the Rosicrucians, of which Descartes and Locke perhaps were members. Newton was not a sceptic; yet his case demonstrates the position typical of philosophers and scientists of the period, in which two simultaneous lines were running in parallel – the metaphysical and the epistemological. The sceptics inclined to fideism based on the inexplicability of God and his influence in the world. Their modest epistemological ambitions, much more modest than those of their non-sceptical colleagues such as Newton or Boyle, combined both the Pyrrhonian humility resulting from the weakness of reason and the humility inherent in fideism.[11]

This was a short but fascinating period in natural philosophy; it was still open to the mysteries and secrets of Nature, enchanted by an infinite universe pulsing with

[10] For details see L. Principe (2000, 2011) and A.G. Debus (2004).

[11] An interesting discussion of the link between Gassendi's fideism and scepticism can be found in S. Murr (1993).

energies, spirits, vapours and mysterious powers that were not directly observable, such as aether, sometimes called the fifth element or the quintessence of the universe. The emerging modern science at the time of Galilei, Gassendi, Boyle and Newton was unique not only for its scientific discoveries but for the strange *mix of rational, sceptical, theological and alchemical influences*; the legacy of the past, including old myths, religious and esoteric elements, generated an atmosphere in which natural scientists still believed in an underlying unity of all being, a unity not demonstrable by empirical science. This situation did not last long: as science advanced, specialized and secularized, the concern of scientists moved away from these spheres; perhaps one could say that science then literally lost its magic. In Hume's time an era was finally coming to a close. Hume left this legacy behind and stayed firmly within the human domain. His epistemological scepticism had no undeclared backing of any metaphysical form; to overcome its destructive consequences Hume had to look for a remedy somewhere *within the human province* – and found it in the spontaneous instincts of human nature. This remedy, however, brought to being a new set of philosophical problems.

References

Annas, J., and J. Barnes. 1985. *The modes of scepticism: Ancient texts and modern interpretations*. Cambridge: Cambridge University Press.

Bayle, P. 1991 [1702]. *Historical and critical dictionary. Selections*, trans. and ed. R. Popkin. Indianapolis/Cambridge: Hackett.

Bourdin, J. 1934 [1641]. The seventh set of objections. In *Descartes' philosophical writings*. Trans. E.S. Haldane and G.R.T. Ross, Vol. II, 259–345. Cambridge: University Press.

Brandt Bolton, M. 1983. Locke and Pyrrhonism: The doctrine of primary and secondary qualities. In ed. M. Burnyeat, 353–357.

Burnyeat, M. (ed.). 1983. *The skeptical tradition*. Berkeley: University of California Press.

Debus, A.G. 2004. *Alchemy and early modern chemistry*. London: Jeremy Mills Publishing.

Descartes, R. 1934a [1641]. Reply to second objections. In *Descartes' philosophical writings*. Trans. E.S. Haldane. and G.R.T. Ross, Vol. II, 30–52. Cambridge: University Press.

Descartes, R. 1934b [1641]. Reply to fifth objections. In *Descartes' philosophical writings*. Trans. E.S. Haldane and G.R.T. Ross, Vol. II, 204–234. Cambridge: University Press.

Descartes, R. 1979 [1641]. Meditations on first philosophy. In *Descartes' philosophical writings*. Trans. E. Anscombe and P.T. Geach, 59–125. Nelson's University Paperbacks.

Dobbs, B.J.T. 2002. *The Janus faces of genius: The role of alchemy in Newton's thoughts*. Cambridge: Cambridge University Press.

Fogelin, R. 1983. The tendency of Hume's scepticism. In ed. M. Burnyeat, 397–412.

Fogelin, R. 1994. *Pyrrhonian reflections on knowledge and justification*. New York: Oxford University Press.

Fogelin, R.J. 2004. The skeptics are coming! The skeptics are coming! In ed. W. Sinnott-Armstrong, 161–174.

Fogelin, R. 1985. *Hume's skepticism in the treatise of human nature*. London: Routledge and Kegan Paul.

Gassendi, P. 1934 [1641]. The fifth set of objections. In *Descartes' philosophical writings*. Trans. E.S. Haldane and G.R.T. Ross, Vol. II, 134–204. Cambridge: University Press.

Hankinson, R.J. 1998. *The sceptics*. London & New York: Routledge.

Hobbes, T. 1934 [1641]. The third set of objections. In *Descartes' philosophical writings*. Trans. E.S. Haldane and G.R.T. Ross, Vol. II, 60–79. Cambridge: University Press.
Huet, P.D. 1725 [1723]. *A philosophical treatise concerning the weakness of human understanding*. London: Gysbert Dommer.
Kemp Smith, N. 2005 [1941]. *The philosophy of David Hume*. Palgrave Macmillan.
Keynes, J.M. 1972 [1944]. Newton the Man. In *Essays in biography*, ed. J.M. Keynes, 363–374. Cambridge: Cambridge University Press.
Leibniz, G.W. 1717. *Leibniz-Clarke correspondence of 1715–1716*. Leibniz' first paper. London. http://www.newtonproject.sussex.ac.uk/view/texts/normalized/THEM00226. Accessed 28 July 2015.
Lennon, T.M. 1993. *The battle of the gods and giants: The legacies of Descartes and Gassendi*. Princeton: Princeton University Press.
Locke, J. 2011 [1689]. *An essay concerning human understanding*. ed. P.H. Nidditch. Oxford: Oxford University Press.
Machuca, D. (ed.). 2011. *Pyrrhonism in ancient, modern, and contemporary philosophy*. Dordrecht: Springer.
Maia Neto, J.R., et al. 2009. *Skepticism in the modern age: Building on the work of Richard Popkin*. Leiden: Brill.
Mersenne, M. 1934 [1641]. The second set of objections. In *Descartes' philosophical writings*. Trans. E.S. Haldane and G.R.T. Ross, Vol. II, 24–30. Cambridge: University Press.
Michael, F.S., and E. Michael. 1990. The theory of ideas in Gassendi and Locke. *Journal of the History of Ideas* 51(3): 379–399.
Montaigne, M. 1993 [1580]. *An apology for Raymond Sebond*. Trans. M.A. Screech. London: Penguin.
Morrison, T. 2011. *Isaac Newton's Temple of Solomon and his reconstruction of sacred architecture*. Basel: Birkhäuser.
Murr, S. 1993. *Gassendi's scepticism as a religious attitude*. Leiden: Brill.
Newton, I. 1934 [1713] *Sir Isaac Newton's mathematical principles of natural philosophy*. Trans. A. Motte, rev. by F. Cajori. Berkeley: University of California Press. https://isaacnewtonstheology.files.wordpress.com/2013/06/theology-of-general-scholium.pdf. Accessed 9 Dec 2015.
Paganini, G. 2003. *The return of scepticism. From Hobbes and Descartes to Bayle*. Dordrecht: Kluwer.
Penelhum, T. 2000. *Themes in Hume. The self, the will, religion*. Oxford: Clarendon.
Popkin, R.H. 1993. *The high road to Pyrrhonism*. Indianapolis: Hackett Publishing Company.
Popkin, R.H. 2003. *The history of scepticism. From Savonarola to Bayle*. Oxford: Oxford University Press.
Principe, L. 2000. *The aspiring adept, Robert Boyle and his alchemical quest*. Princeton: Princeton University Press.
Principe, L. 2011, January 6. The sceptical chymist. *Nature* 469: 30–31.
Schmitt, C.B. 1983. The rediscovery of ancient scepticism in modern times. In ed. M. Burnyeat, 225–253.
Sinnott-Armstrong, W. (ed.). 2004. *Pyrrhonian skepticism*. New York: Oxford University Press.
Snobelen, S.D. 2001. '*God of gods, and Lord of lords': The theology of Isaac Newton's* General Scholium *to the* Principia.
Vázques, B. 2015. *The scepticism of Michel de Montaigne*. Cham: Springer.
Walker, R. 1983. Gassendi and skepticism. In ed. M. Burnyeat, 319–337.
Westfall, R.S. 1982. Newton's theological manuscripts. In *Contemporary Newtonian research*, Studies in the history of modern science, ed. Z. Bechler, 129–143. Dordrecht: Reidel.
Williams, B. 1983. Descartes' use of skepticism. In ed. M. Burnyeat, 337–353.
Woolhouse, R.S. 1993. *Descartes, Spinoza, Leibniz. The concept of substance in seventeenth-century metaphysics*. London: Routledge.

Chapter 4
The Pyrrhonian Roots of Hume's Scepticism

Abstract A strong influence of Pyrrhonism on Hume's thought explains why he so easily accepted the weakness of reason as a predicament we have to live with. Hume's corrected (for some, perfected) version of Pyrrhonism follows on the sceptical principle that we cannot see behind appearances, but newly asserts the power of natural inclinations in life. Hume's principal correction of Pyrrhonism consists not in diluting radical (excessive) scepticism in epistemology, but in allotting intense reflection and instinctive beliefs their own domains. Hume daringly asserted a disparity that involves a division of labour between reason and nature in place and time. In contrast to the old Pyrrhonism Hume set out to redefine *ataraxia* from a state of life lethargy to embracing life with all its pleasures. Hume teaches how to be happy, moral, socially active citizens and how to promote tolerance in all these areas of human behaviour. It is in common life, in moral and social areas, that mitigated scepticism has its role – it put us in a pleasant frame of mind, makes us feel agreeable and cultivates calm passions.

Keywords Excessive scepticism • Epistemology • Common life • Morality • Happiness

4.1 The Force of Hume's Scepticism

Hume absorbed all the diverse influences of this turbulent atmosphere and was thus stimulated to form his own views. He (most probably) read the original texts of Diogenes and Sextus Empiricus.[1] As did all philosophers of the time, he took part in disputes concerning Cartesianism. These disputes, discussed in the previous chapters, focused on Descartes' conception of science, mind and metaphysics: Descartes' criticism of the Aristotelian conception of nature was widely acknowledged; the response to his purely intellectual definition of mind (and to the subjective grounding of cognitive certainty) was mainly negative; and Descartes' version of metaphysics met with a mixed reaction. Hume ignored Descartes' natural philosophy

[1] Peter Fosl discusses in detail the historical questions of Hume's access to and use of the original sources of ancient scepticism (1998; 2011).

Z. Parusniková, *David Hume, Sceptic*, SpringerBriefs in Philosophy,
DOI 10.1007/978-3-319-43794-1_4

and rejected his metaphysics. As for the second domain, he picked up the most radical theme consisting in the autonomy of mind (the recipient of perceptions and the centre of thought operations) with no attachment to any external authority; he faced the consequences of this position's leading to scepticism about any existences independent of the mind. Descartes' turn to autonomous subjectivity was a constitutive factor in the change from ancient metaphysics and scholastics to modernity; Hume played an important role in this process. Further, Hume rejected Descartes' conception of innate ideas and endorsed experimental method in epistemology.

Hume was also influenced by the group of constructive sceptics with whom he shared the empiricist position. He was inspired by their Pyrrhonian attitude, specifically by their acceptance of the weakness of reason, resulting in modesty and caution in judgment, in an emphasis on appearances and on tolerance towards other views. However, Hume recognized an inconsistency in their attitude to knowledge in which they limited themselves to making only a few concessions to Pyrrhonism: they replaced the ideal of the certainty of knowledge with probability; and both their scientific and fideist orientation prevented them from adopting a full scale phenomenalism.

In this respect, they assumed the position of the ancient Pyrrhonians, as expressed by both Diogenes and Sextus. Diogenes says, "I do not lay it down that honey is sweet, but I admit that it appears to be so"; and similarly Sextus: "it appears to us that honey sweetens but we cannot assert that it is actually sweet" (*PH* I, 19, 8; *DL* IX, 105, 517). The constructive sceptics would not doubt the existence of honey, only our ability to assert the truth about it; but unlike the ancient sceptics, who limited knowledge strictly to appearances, they claimed to have probable knowledge about reality. This position may resemble what Fogelin terms the "milk and water scepticism" (1985, 146) falsely ascribed to Hume, specifically to his concept of mitigated scepticism; the uniqueness of Hume's position consists in the fact that his epistemological scepticism remains unmitigated and yet his epistemology does not have a destructive, paralyzing effect.

This line of interpretation takes us to Pierre Bayle and his direct and decisive impact on Hume's philosophy. Mossner remarks that before Hume left for La Fleche more than 50 % of his notes and memoranda were on Bayle; he packed into his luggage, when he eventually left for La Fleche to write his *Treatise*, Bayle's *Dictionary* (*Dictionnaire historique et critique*, 1702), totalling more than six million words. Bayle was a proponent of the radical, destructive Pyrrhonism; he never created any philosophical system of his own but whichever new system appeared he began to undermine it as a matter of principle, exposing its dogmatic features. Bayle shows the hopelessness of solving the sceptical contradictions and in this respect his scepticism displays the same intensity as Montaigne's. There is no middle way – say, in probability – between scepticism and true knowledge. To repeat Montaigne's words, "we judge entirely or entirely not" (Montaigne 1993, 141); here Montaigne exposes the sceptical dilemma crystal clear, following Pyrrho's message: "for if certain things only are true <and others are false>, how are we to distinguish between them? Not by senses, where things in the field of sense are in question, since all these things appear to sense on equal footing; nor by the mind for the same reason" (*DL* 1925, 505).

Bayle follows this radical line and says, "our reason is only suitable for making everything perplexing and for raising doubts about everything. No sooner has it built something than it provides means for destroying it" (Bayle 1991, 42, Bunel, rem. E); and in a similar tone, our reason "is a principle of destruction and not of edification. It is only proper for raising doubts, and for turning things on all sides in order to make dispute endless … It is only fit to make man aware of his own blindness and weakness" (Bayle 1991, 151, Manicheans, rem. D). Bayle was the first philosopher to explicitly attack the distinction between primary and secondary qualities; in the article on Pyrrho, remark B, he says, "for if the objects of our senses appear colored, hot, cold, odoriferous, and they are not so, why can they not appear extended and shaped, in rest and in motion, though they are not so?" (Bayle 1991, 197). Bayle anticipated Hume's criticism of Locke and his defence of primary qualities against scepticism. As Hume repeatedly says, "if all the qualities, perceived by the senses, be in the mind, not in the object, the same conclusion must reach the idea of extension, which is wholly dependent on the sensible ideas or the ideas of the secondary qualities" (E 154).[2]

The result of Pyrrhonian scepticism is devastating for our state of mind and, as a true Pyrrhonian, Bayle must have been concerned about it – like Montaigne before him and Hume after. The closing passages of Hume's *Treatise* and, to a lesser extent, of his *Enquiry* contain lamentations about the sufferings we have to undergo when confronted with Pyrrhonian contradictions; scepticism makes us desolate instead of enabling our minds' repose. Unlike the constructive sceptics, who did not pay attention to this aspect of Pyrrhonism, science being of the utmost importance to them, the true Pyrrhonians, such as Bayle and Hume, focused upon a *balanced, quiet or even joyful state of mind as the main goal of philosophy*. However, in the modern atmosphere, with epistemology becoming the '*philosophia prima*', the impotence of reason was more disconcerting than in ancient times and the mere awareness of this prevented man from achieving that serene state. Scientists, as we could see, avoided the total suspension of judgment by compromising the integrity of the Pyrrhonian tropes. Montaigne, Bayle and Hume held to the logic of Pyrrhonism and admitted that suspension of judgment was unavoidable in matters of cognition; yet they felt the need for some positive anchor in life, for something more than just the passive, traditional conformity which the Pyrrhonians advised. Montaigne and Bayle found it in fideism, Hume in human nature. In his views on scepticism, Hume followed the radical line that runs from Montaigne and Bayle, not the meagre mitigated scepticism of the constructive empiricists, though the common ground in empiricism may tempt us to perceive such a link.

Anyway, Bayle's fideism has a desperate urgency that corresponds with the intensity of his scepticism. There he found a private refuge from the destructive

[2] Gassendi did not make this distinction and could thus be seen as more faithful to the Pyrrhonian ideal that we cannot get beyond the appearances of *things*; but he proposed a similar distinction between the more permanent qualities – magnitude, size and shape – and the fleeting sensible qualities; still, Gassendi's basic assumption is that all qualities are just compounds of the same essential particles of matter, atoms.

effects of scepticism; it provided for him an "impenetrable shield against the arrows of the Pyrrhonists" (Bayle 1991, 196). Pyrrhonism is, according to Bayle, the basis for Christianity: "a man is therefore happily disposed toward faith when he knows how defective reason is" (Bayle 1991, 206). Bayle draws directly on Montaigne in his references to St. Paul[3] and says that "those who have always professed humility and ignorance accommodate themselves much better than others to this spiritual darkness [into which God retreated]" (Bayle 1991, 205). Scepticism is said "to be of great value to the Christian soul since it makes him give up those doctrinaire opinions that St. Paul detested so strongly" (Bayle 1991, 205).[4] As mentioned before, one can find many parallels between Bayle's fideism and the position of Demea in Hume's *Dialogues*; scepticism is no threat to faith since faith is beyond reason. Bayle's view that giving up reason is "a great step toward the Christian religion" (Bayle 1991, 206) and his observation that "[Pyrrhonism] … can most docilely accept the mysteries of our religion" (Bayle 1991, 205) resonate with Demea's statements about "the adorable mysteriousness of the divine nature" (Hume 1998, 48). Bayle's Third Clarification on Pyrrhonism (an appendix to the *Dictionary*) is completely devoted to fideism and can be summarized by this poetic quote: "The bark of Jesus Christ is not made for sailing in this stormy sea [of theological disputes], but for taking shelter from this tempest in the haven of faith" (Bayle 1991, 423).

It is interesting to note that the fideism of the constructive sceptics did not contain such exulted irrationalism – their situation was not felt as desperate as to posit so sharply the contradiction between reason and faith. It may seem at first sight that Huet speaks in Bayle's language when he claims that man "should willingly submit himself to the Mysteries of Faith, which are obscure, and so much above our Senses and Reason" (Huet 1725, 177). But for him faith stands above – not against – reason such as in Bayle's case. Huet adopts a soberer, less mystical version of fideism. Although our faith is due to the grace of God and his inward light, God nevertheless "comes to help and support the Imbecility of our Nature and of our Reason" and thus seems to work in tandem with reason, for "Reason stands in need of this Assistance of the divine Grace" (Huet 1725, 218).

For Hume, fideism or any other religious dimension lost relevance as a solution of scepticism. However, he sympathized with Bayle's fierce criticism of the Church and its oppressive practices, for which Bayle had been labelled "the arsenal of the Enlightenment". In his early but widely-read work of 1682, *Pensées diverses sur la comète*,[5] Bayle argues that a society of atheists would be more stable, peaceful and prosperous; social laws have their origins in the human sense of justice, not in

[3] Montaigne refers to St. Paul's First epistle to the Corinthians, for instance 1:20 (Where is the wise? Where is the scribe? Where is the disputer of this world? Hath not God made foolish the wisdom of this world?), and 3:20 (And again: The Lord knoweth the thoughts of the wise, that they are vain.). This reference is also used by Huet (1725, 19).

[4] Bayle refers to the views of a well-known sceptic, La Mothe le Vayer.

[5] This strange-sounding title refers to the panic and various miracle theories arising from the appearance of the Great Comet in 1680; Bayle supported a scientific explanation.

ancient or sacred texts: "human justice constitutes the virtue of the majority of the world" (Bayle 2000, 200). This attitude appealed to Hume, who shared this Enlightenment spirit (although he was not an outspoken atheist) and made it even more consistent by having left the fideist 'baggage' behind. The counterbalance to scepticism must, for Hume, come from the human province alone. It is nature which breaks the chain of sceptical reasoning and transports us to the domain of instinctive behaviour and common sense; here our scepticism evaporates like smoke. Thus nature, not faith, saves us from this agony we suffer in the closet. Natural, unconditional instinctive assent concerns our basic beliefs in the existence of bodies (external objects) and causality, and our expectations that the future will resemble the past.

This domain of common sense that cannot reflect upon itself guarantees our survival; in my native language, Czech, common sense literally translates as "healthy reason" – a perfect counterpart to the sceptical reason that would paralyze even our survival instincts were it not for the sweeping power of nature. Part of the survival strategy is, of course, knowledge; we have natural cognitive instincts that we naturally – and successfully – use in life. Hume does not compromise on epistemological doubts that are the necessary result of sceptical arguments; yet our instincts convey the opposite message. Does this not mean that we live in two separate worlds, the world of reason and the world of instincts, each having its own rules, norms and semantics? Are we not torn between these two worlds suffering from some kind of schizophrenia? Should we not stop sceptically undermining our natural ways of understanding because scepticism brings only pain? Is there any suggestion in Hume's writings of reconciliation between these two powers? These are the key questions when assessing Hume's scepticism.

Hume's own version of Pyrrhonism unfolds from a tension that he identified between the ancient Pyrrhonian injunction to suspend judgment and his own proposition that beliefs must be embraced without any second (or, rather, first) thought. Basic instincts are unshakable for philosophers and ordinary folk alike; none of them, not even informed philosophers, can be made to feel that the real existence of the tree we see in the garden is doubtful and may be just a sequence of fleeting sensations. Moreover, these instincts are shared by all living creatures: "even the animal creation are governed by a like opinion, and preserves this belief of external objects, in all their thoughts, designs, and actions" (E 151). This attitude, however, is in conflict with the ancient Pyrrhonian injunction to suspend judgment on anything but appearances.

Hume recognized that the ancient Pyrrhonians required us not only to suspend theoretical judgement as a result of sceptical considerations but also *to withhold, or at least to suppress, our naturally affirmative beliefs and to deny the epistemic value of appearances*. For Hume, the Pyrrhonian position assumed that we would *indifferently* report on appearances; but this would require the exercise of *impossible restraint* on our nature. This requirement *is an example of excessive scepticism that would destroy life*: "all discourse, all action would immediately cease; and men remain in a total lethargy, till the necessities of nature, unsatisfied, put an end to their miserable existence" (E 160). In their crusade against dogmatism the old

Pyrrhonians insisted on a denial of strongly felt (dogmatic) convictions. On the one hand, this instruction is ironically *dogmatic*, breaking the rule of being non-committal and pushing man to violate his nature; on the other hand, it is impossible to fulfil since man cannot be rid of his animal dimension and torn from the realm of nature. Hume set out to correct this contradiction by offering a solution that consisted in separating the suspension of judgment from the suspension of belief. In this move he rehabilitates the domain of our nature and shows that it is beyond the reach of sceptical scrutiny.

4.2 Did Hume Interpret Pyrrhonism Correctly?

Before looking at Hume's solution let us consider if the Pyrrhonians really wanted to suspend beliefs. Did Hume not misinterpret their scepticism by attributing to it a tendency to *live this extreme scepticism*? It is difficult to provide univocal answers to these questions, given the limited original sources: Pyrrho (360–270 BC), for instance, did not write anything. By the way, despite writing nothing, Pyrrho was admired by Epicurus and the city of Elis made him a high priest: "on his account they [the citizens] voted that all philosophers should be exempt from taxation" (*DL* IX, 64, 477) – an amazing achievement not matched since by any philosopher. Pyrrho's teachings were developed by a number of philosophers, by Pyrrho's adherent Timon (app. 325–230 BC) and, after a lengthy break, by Aenesidemus and Agrippa (first/second centuries AD), and recorded in a brief version by Diogenes Laertius (c. third century AD) and by Sextus Empiricus (160–210 AD). These two are the main sources for the research of Pyrrhonism.

A good outline of the problem of the Pyrrhonian attitude to beliefs is presented by Myles Burnyeat in the essay, "Can the Sceptic Live his Scepticism?" (1998), in which he links his interpretation of Pyrrhonism to Hume (Burnyeat and Frede 1998). Burnyeat argues in favour of Popkin's version of Hume's account of Pyrrhonism that consists in attributing to Pyrrhonism the tendency to suspend our beliefs and to live by that result. Sextus defines scepticism as follows: "the chief constitutive principle of scepticism is the claim that to every account an equal account is opposed; for it is from this, we think, that we come to hold no beliefs" (*PH* I, 12, 6). Sextus also repeatedly emphasizes that we should live by appearances, "in accordance with everyday observances without holding opinions" (*PH* I, 23, 9). Burnyeat points out that living by appearances contradicts the life of belief. He then looks at Sextus' description of the ideal Pyrrhonian way of life. The recipe is four-fold and consist in (a) guidance by nature, (b) necessitation by feelings, (c) handing down of laws and customs and (d) teaching some form of expertise (*PH* I, 23, 9). It may seem that (a) and (b) support the opposite interpretation in which the Pyrrhonians accept natural feelings without reservation. Yet (a) only means that we are naturally equipped with the ability to perceive and think and (b) refers to the bodily drives enabling our survival that have nothing to do with belief; (c) is as the moral ability to distinguish between good and evil, piety and impiety and (d) is a recommendation to practice one's art or profession, medicine in Sextus' case. None of the four recommendations concerns belief.

If we look at Diogenes' recording of Pyrrho, the tendency to detach oneself from belief is more visible. Pyrrho does not appeal to any serious practice of a profession, or to a civil involvement in life and society; he rather reinforces the isolationist tendencies. After all, when he became bored with his pupils' questions during a session that took place near the river Alpheus, he stripped and swam across it (*DL* IX, 69, 481). The 'Pyrrhonian principle' (according to Aenesidemus) consists only in a "report on phenomena or on any kind of judgement, a report in which all things are brought to bear on one another, and in the comparison are found to present much anomaly and confusion. As to the contradictions in their doubts, they would first show the ways in which things gain credence, and then by the same methods they would destroy belief in them" (*DL* IX, 78, 491).

Diogenes also tells us various anecdotes from Pyrrho's life that confirm his detached attitude: "when septic salves and surgical and caustic remedies were applied to a wound he had sustained, he did not so much as frown" (*DL* IX, 67, 479); "he showed his indifference by washing a porker" (*DL* IX, 67, 479); and a story in which Pyrrho travels on a ship in a storm and calms the unnerved passengers, "pointing to a little pig in the ship that went on eating, and telling them that such was the unperturbed state in which the wise man should keep himself" (*DL* IX, 68, 481). At moments, his indifference bordered on cruelty: "when Anaxarchus fell into a slough, he passed by without giving any help, and, while others blamed him, Anaxarchus himself praised his indifference and *sang-froid*" (*DL* IX, 63, 479). These examples show that for the Greeks, *bios theoretikos* played a dominant role, being often superimposed upon practical life and beliefs. Bayle emphasized that Pyrrho's *indifference* resulted from the suspension of both judgment and belief. This indifference even suppressed Pyrrho's self-preservation instinct.[6] He did not show positive feelings nor was he ever angry; he did not care who attended his speeches and continued speaking even if his audience went away; he put equality between life and death; and, most importantly, "he especially despised human nature, and he was forever repeating the words of Homer, where he compares men to leaves" (Bayle 1991, 208). Hume was an avid reader of Bayle and could have been inspired by him in this interpretation of the Pyrrhonians. In this context, Myles Burnyeat asks the question which was, as he claims, crucial for Hume: "why this should produce tranquility rather than acute anxiety?" (Burnyeat 1998, 55). Hume rightly diagnosed the psychological agony caused by this attitude that effectively prevents man from achieving the ultimate Pyrrhonian goal, *ataraxia*. Hume's position is that *human life is not possible without beliefs entailing epistemic commitments; but then, any efforts to avoid beliefs would necessarily cause agony instead of tranquility*.

Michael Frede takes a different approach. The primary aim of the Pyrrhonians was apparently to expose the dogmatists' claim to have deeper insights into the true nature of things. Frede claims that it is wrong, as Burnyeat does, to ascribe to the

[6] "Neither a chariot nor a precipice could ever make him take a step forward or backward … his friends who followed him around often saved his life" (Bayle 1991, 195).

Pyrrhonians the demand that one live without beliefs; a suspension of judgment is not a suspension of belief, involving not assenting to any proposition. One can find some textual evidence supporting this interpretation. Sextus defines belief as follows: "when we say that Sceptics do not hold beliefs, we do not take 'belief' in the sense in which some say, quite generally, that belief is acquiescing in something; for Sceptics assent to the feelings forced upon them by appearances" (*PH* I, 13, 6). Sextus clarifies this point further: "We say that they do not hold beliefs in the sense in which some say that belief is assent to some unclear object under investigation" (*PH* I, 13, 6). In any case, the criticism of the position asserting that we may have access to the utmost secrets of nature also played an important role in Hume's scepticism.

Frede (1998a, b) argues that a sceptic cannot avoid thinking about himself as knowing something and thus cannot expel beliefs; however, Frede seems to intellectualize belief and treat it as some result of reflection; he seems to think that if a sceptical reflection taught us to be satisfied with appearances, perhaps the satisfaction with appearances may work in the case of beliefs, too. Frede distinguishes between wider and narrower belief, the former referring to how things are and the latter to thinking how they appear; only the former is dogmatic and should be suspended. Frede's conclusion is that if the sceptic suspends judgment on how things are, that "by no means rules out that he should have beliefs about how things appear to him" (Frede 1998a, 9). This conclusion seems to contradict the very meaning of belief; the word 'belief' comes from the Greek word '*dogma*'. Thus belief cannot refer to appearances that are only provisional subjective records of our perceptions but in fact concerns reality and truth.

Jonathan Barnes (1998) presents an interesting interpretation in which he claims that Pyrrhonism is no general orthodoxy. According to him, Pyrrhonians work piecemeal and exercise scepticism on particular issues (as Hume sometimes also suggests). Barnes distinguishes between rustic scepticism (adopted, as he argues, by Burnyeat), and urbane scepticism (favoured by Garrett); according to urbane scepticism we can hold beliefs about ordinary things and ordinary causes of events. According to this position, the sceptics conform to *bios* as it is lived and not to the prescriptive general norms considered superior to this life. Barnes refers to Diogenes' words: "we also perceive that fire burns; as to whether it is its nature to burn, we suspend our judgement" (*DL* IX, 104, 515); and he shows that the ancient sceptics wanted merely to suspend judgment on the 'ultimate principles' of particular observable processes. No doubt this motive influenced modern empiricism, including Hume. However, did the Pyrrhonian sceptical suspense not extend further? Diogenes' words refer to judgment; but suspending judgment does not affect the power of our belief in the truth of the statement that fire always burns. On the whole, the stories from Pyrrho's life show that he tried to apply detachment from *bios*, suppressing not only beliefs but even spontaneous survival instincts, such as pain and hunger. Barnes wonders whether the sceptics could have avoided total inaction had they resisted beliefs – a good question considering Pyrrho; but he was, as the story goes, often helped by his pupils to get along with life. Moreover, both Pyrrho and Timon lived to their nineties and Timon was especially active – he was a dancer, a poet, a playwright and a lover of wine!

Hankinson supports some claims made by Burnyeat. He argues that "what distinguishes the Sceptical state of mind from any other is its lack of commitment to any truth … the Sceptic will back no horses. He may have views, but … has nothing invested in them. They will blow away at the first puff of a phenomenal wind" (Hankinson 1998, 94–5). He considers legitimate Burnyeat's concerns that the sceptic's life without beliefs produces extreme anxiety, not tranquility. Perhaps, he says, the sceptic's 'assent' is just limited to conditioned reflexes; that may be the guidance of nature mentioned by Sextus, involving instincts necessary for survival; this could be the way of interpreting Diogenes' remark that "the Sceptic replies that he will be able so to live as to suspend his judgement in cases where it is a question of arriving at the truth, but not in matters of life and the taking of precautions" (*DL* IX, 108, 519).

In that case, belief would have an entirely passive form – *it will not react contrary to appearances*; it corresponds to Sextus' advice to display arguments for and against some proposition, weigh them and declare a draw; we have no choice than to suspend judgment. In this case, the lack of belief would paradoxically not end in anxiety, Hankinson says. True, the scope of Pyrrhonian doubt is universal, not sparing anything, but due to the passive attitude *the sceptic's position is always provisional, unattached*, while anxiety involves some concern for the way things are going to turn out. The sceptic is indifferent to any threat that some argument may emerge to disturb the status quo. Hankinson introduces a picture of a sceptic as a 'zombie' floating through life without belief and without epistemic commitments, in the state of permanent *ataraxia*. This may be partly applicable in Pyrrho's case, illustrated for instance by this anecdote from his life: "when a cur rushed at him and terrified him, he answered his critic that it was not easy entirely to strip oneself of humanity" (*DL* IX, 66, 479). It seems that Pyrrho tried his best to do that. Hume, in Burnyeat's view, showed the impossibility and *the absurdity* of so doing: "when one has seen how radically the sceptic must detach himself from himself, one will agree that the supposed life without belief is not, after all, a possible life for man" (Burnyeat 1998, 57). Moreover, 'unhuman' life (without belief) should not be seen as the goal of scepticism.

In any case, the question about the Pyrrhonian implementation of scepticism in life remains open. However, the interpretation of this problem by Bayle and Montaigne could have influenced Hume (and other modern sceptics) in his understanding of Pyrrhonism. Montaigne argues that "the Sceptics keep their assent in a state of ambiguity, inclining neither way, giving not even the slightest approbation to one side or the other" (Montaigne 1993, 70). Their attitude goes to the extreme since they do not expect anybody to believe them either: "when they assert that heavy things tend to follow downward, they would be most upset if you believed them" (Montaigne 1993, 70). While Hume rehabilitates belief as an independent – and even stronger – faculty of our nature, a faculty that *necessarily gives assent*, Montaigne's interpretations of Pyrrhonism emphasize the reverse: "[the soul] bestows assent on nothing. This leads to their well-known *ataraxia*: that is a calm, stable rule of life, free from all the disturbances which give birth to fear, avarice, envy, immoderate desires, ambition, pride, superstition, love of novelty, rebellion, disobedience, obstinacy and the greater part of bodily ills" (Montaigne 1993, 70). To get rid of these violent passions we have to shake off all convictions and suspend spontaneous beliefs. This was the point of departure for Hume's criticism of Pyrrhonism.

However, this passage from Montaigne demonstrates another interesting motive that affected Hume. Scepticism – and our reaction to it – was not a purely epistemological matter but concerned the frame of mind and the whole of life; both scepticism and *ataraxia* influence, in opposite ways, our psychology and are connected with emotions, even with our health (see Appendix 2). Modern interpretations of scepticism often concentrate on its dangers in undermining all human cognitive capacities; ancient Pyrrhonism is often reduced to issues of epistemology, as discussed in the sceptical tropes. However, this is an anachronism that transfers the modern interest in knowledge to a different discourse. The ancient approach was different; doubting (and the escape from it) was *a way of life*, not just an epistemological *exercise*. Diogenes ascribes to 'Pyrrhonians' four characteristics: they are "The Zetetics [*zeteo* – I seek] or seekers because they were ever seeking truth, Sceptics [*skeptesthai* – to reflect] or inquirers because they were always looking for a solution and never finding one, Ephectics [*epechein* – hold back] or doubters because of the state of mind which followed their inquiry, I mean, suspense of judgement, and finally Aporetics [*aporos* – without passage] or those in perplexity, for not only they but even the dogmatic philosophers themselves in their turn were often perplexed" (*DL* IX, 69–70, 483). If we swap around the last two characteristics, the way through life of a person whom we commonly call a sceptic[7] begins by seeking the truth and making inquiries about specific subjects; since against every account there stands another without any possibility of deciding between them, we are left perplexed; finally, when doubts are resolved by the suspension of judgment we enter into the state of *ataraxia*. Ancient scepticism is not just a theoretical procedure but involves our whole being and affects our entire life.

The anachronistic modern emphasis on the epistemological impact of scepticism encourages the view that the ancient sceptics demanded only a suspension of judgment. However, Hankinson and Burnyeat point out that Pyrrhonian scepticism affects both our everyday perceptions and theoretical constructions: our beliefs and reflections. *Pyrrhonism does not practise insulation by subject-matter but it is an argumentative practice and a way of life*. Pyrrhonism weakens our attachment to life and subdues the *feeling of urgency of our beliefs*. Yet on the intellectual level, the Pyrrhonians are seekers and inquirers, searching in vain for truth, working hard to detect and expose any views offered by the dogmatists. But on the emotional and practical levels, Pyrrhonians are *passive*; the term 'ataraxia' is derived from '*tarassein*' (disturb) and the prefix 'a', forming the word '*ataraktos*'. The way to ataraxia

[7] The basic definition of scepticism given by Sextus Empiricus mentions the ability to set out oppositions among things, and whilst finding that none of the conflicting account can be preferred, we suspend judgment and achieve tranquillity (*PH I* 8–10, 4–5).

is by practicing *apaheia*, freedom from passions. They minimize any attachment to beliefs and any passionate involvement in life; their tranquility is serene and *quiet*.

Hume stands in opposition to Pyrrhonian lethargy and shares the spirit of modernity that is active, optimistic, charged with energy. Hume set out to redefine *ataraxia* from a lifeless state to embracing life with all its pleasures. In contrast to the Pyrrhonian 'antiseptic' detachment from life, Hume encouraged man to enjoy life to the full, to follow feelings, beliefs, desires, to get involved in a community, in trade and business, political matters, to converse with friends, play backgammon and drink wine. For Hume (as for the Pyrrhonians), *philosophy has a practical mission*; but unlike the Pyrrhonians, Hume *teaches how to be happy, moral, socially active citizens and how to promote tolerance in all these areas of human behaviour*. For Hume, the *Pyrrhonian apathy is not a therapy but a disease*. He showed that it is not possible, as Sextus demanded, "to live by experience and without opinions" (*PH* II, 246, 136) and that it is not desirable to recommend the ideal of the sceptics consisting in "moderation of feelings in matters forced upon them" (*PH* I, 29–30, 11). However, Hume also acknowledged that Pyrrhonian scepticism is unbeatable by argument; sceptical reflection on epistemological issues has *no cure within its own argumentative domain*.

4.3 Hume's Correction of Pyrrhonism

Hume's specific conception of scepticism unfolds from his rejection of the Pyrrhonian suspension of belief. Hume isolated the area of epistemological scepticism from beliefs because he believed that this division ensured a positive attitude to both knowledge and the world. Judged rationally, our knowledge is produced by instinctive procedures in our minds and thus it has no rational justification. Yet instincts have the force to carry us away and to form positive unshakable beliefs. The domain of beliefs cannot be affected by scepticism because it has a sweeping primal power immune to any kind of sceptical reflection. Hume discovered an inconsistency in the Pyrrhonian claim that *we must both stick to appearances and simultaneously suspend belief in them*. Pyrrho can claim that we cannot assert that honey is sweet but only that it sweetens in a perceptual way [it appears to be sweet], but we cannot get rid of the overwhelming belief that it is really sweet. Pyrrhonians, unlike Hume, did not acknowledge that despite having no justifiable basis for our beliefs we still have them, and no instruction to distance ourselves from them can lead to peace of mind; it would drive us mad instead.

In sum, the ancient Pyrrhonian asks us to *exorcise any flavour of reality from all naturally emerging convictions* and to practise maximum detachment from life. Hume corrects what he thinks is an agonizing state of mind, resulting from the Pyrrhonian ban on our whole-hearted, positive acceptance of beliefs. Instead, as Richard Popkin argues, Hume encourages the sceptic to apply radical doubts in reflection and at other times to assent to natural beliefs as would any other man. The trick to escape the psychological torture inherent in ancient Pyrrhonism is to *sepa-*

rate the two areas and grant each of them its own domain. Scepticism as a destructive argumentative position cannot and *need not* be mitigated and beliefs can retain their full persuasiveness. Hume has often been accused of creating a clash between reason and nature; after all, sceptical conclusions and beliefs 'tell' us contradictory things. There is no mitigation in the sense of some conciliatory interaction between reason and instinct; As Richard Popkin argues, "The Humean skeptic, the consistent Pyrrhonian, will doubt when he must and believe when he must" (Popkin 1993, 145–6). The rational and the natural keep their own truths in full, the former deadly sceptical, and the latter sweepingly positive.

However, does this situation not lead to even greater pain than the original Pyrrhonian suspension of beliefs? Not really, if we acknowledge that *we do not inhabit the rational and the natural domains simultaneously.* In this move, Hume distances himself from the old Pyrrhonists ('those sceptics') and calls them a 'fantastic sect' that cannot see that "neither I, nor any other person was ever sincerely and *constantly* [emphasis mine] of that [totally sceptical] opinion" (T 183). Similarly, "a Pyrrhonian cannot expect that his philosophy will have any *constant* [emphasis mine] influence on the mind" (E 160). We can question the legitimacy of beliefs but our sceptical investigations will soon be overpowered by the belief itself: "*belief is more properly an act of the sensitive, than of the cogitative part of our natures*" and thus escapes the sceptical arrows of Pyrrhonism (T 183).

Hume's principal correction of Pyrrhonism consists not in diluting radical scepticism to some harmless mitigated sort, but in allotting intense reflection and basic beliefs their own domains. This involves a *division of labour between scepticism and belief in place and time.* For Hume, sceptical arguments show the rational illegitimacy of our reasoning but, metaphorically, these verdicts do not leave the courtroom. Nature is not affected by rational reflection and keeps her own wisdom. Thus, a true Pyrrhonist is both a dogmatic believer and an extreme sceptic, both part-time (although nature takes more of the workload). Unlike the excessive ancient Pyrrhonist who extends the impact of scepticism to the area of beliefs and behaviour, Hume proposes expelling sceptical doubts from our basic beliefs and thus *keeping theoretical scepticism in its most radical form.* This interpretation provides a challenging and original insight into Hume's scepticism, arguing for a divorce of the two powers; the partners are incompatible and they will be better off if separated.

Hume argues that we either carry on with our sceptical reflections, in darkness and dreary solitude, or we carry on with life, in sunshine and company, fully absorbed in one or the other. Reason and instincts are two different worlds that play by different rules and since each displays itself in different circumstances they do not compete or overlap. Instead of being torn apart we accept the dichotomy of our constitution; a sceptic can switch from one world to the other. When feelings and beliefs overwhelm us with irresistible force we simply yield to them; when we practice scepticism we find nothing but doubt and ignorance; this state may be agonizing but does not last long since nature saves us in time. This is the picture of "a perfect Pyrrhonist in his two moods … in one mood, the difficulties overcome him, in another, necessities do" (Popkin 1993, 132). Later (especially at the end of Book

I of the *Treatise*) Hume put more emphasis on the unpleasant feelings caused by profound reflection and seemed keen to abandon it altogether; this inclination, however, did not affect his epistemology but signaled his increasing focus on common life and moral science.

Hume acknowledges the contradictions between the rational (reflective) and the instinctive, be it with regard to causation, induction or external existence. On the one hand there is "the natural propensity of the imagination, to ascribe a continu'd existence to those sensible objects or perception, which we find resembling each other in their interrupted appearance"; on the other hand, "when we compare experiments, and reason a little upon them, we quickly perceive, that the doctrine of the independent existence of our sensible perceptions is contrary to the plainest experience" (T 210). Hume argues that "the smooth passage of the imagination along the ideas of the resembling perceptions makes us ascribe to them a perfect identity. The interrupted manner of their appearance makes us consider them as so many resembling, but still distinct beings, which appear after certain intervals. The perplexity arising from this contradiction produces a propension to unite these broken appearances by the fiction of a continu'd existence" (T 204). Furthermore, "nor have we any idea of *self*, after the manner it is here explain'd. For from what impression cou'd this idea be deriv'd? This question'tis impossible to answer without a manifest contradiction and absurdity;" (T 251). Nor have we any proper idea of causation, except the experienced frequent succession of two separate events; this makes us "*feel* a new sentiment or impression … and we begin to entertain the notion of cause and connexion" (E 78).

We move between these two powers but we cannot, so to speak, stand astride with one foot here and the other there and hope to retain balance; we have to stand firmly either on one or on the other. As Hume notes, "so far, then, are we necessitated by reasoning to contradict or depart from the primary instincts of nature, and to embrace a new system with regard to the evidence of our senses. But here philosophy finds herself extremely embarrassed, when she would justify this new system, and obviate the cavils and objections of the sceptics" (E 152). Meanwhile, "whoever has taken the pains to refute the cavils of this *total* scepticism, has really disputed without an antagonist, and endeavour'd by arguments to establish a faculty, which nature has antecedently implanted in the mind, and render'd unavoidable" (T 183).

In the *Abstract*, Hume – with regard to geometry, but applicable generally – uses even stronger words: "'Twere certainly to be wish'd, that some expedient were fallen upon to reconcile philosophy and common sense, which … have wag'd most cruel wars with each other" (T *Abstract* 659). It is a flashy but somewhat exaggerated expression; our nature is certainly not in any war, it is untouched and undisturbed by our sceptical philosophy – sceptical doubts strive to uproot instincts, though in vain. We are faced with the fact that "such an opinion [of external existence but, again, valid for causation or induction] if rested on natural instinct, is contrary to reason, and if referred to reason, is contrary to natural instinct, and at the same time carries no rational evidence with it" (E 155). Ultimately, man is overwhelmed by "universal perplexity and confusion, which is inherent in human nature" (E 161), and "reason here seems to be thrown into a kind of amazement and suspence" (E 157). These are the effects of profound philosophical objections against the legitimacy of our reasoning that lead us to the conclusion "that nothing leads us to this inference but custom or a certain instinct of our nature" (E 159).

These contradictions, causing the state of amazement, confusion and perplexity (*aporia*), belong to the typical Pyrrhonian vocabulary; Hume's perfection of Pyrrhonism broke the link between *aporia* and *epechein* – perplexity and suspension of judgment. By arguing against attempts to transfer scepticism to belief that operates in the domain of our natural tendencies, in other words by insisting on the isolation of scepticism and limiting it to reflection only, Hume introduced the possibility of accepting this "whimsical condition of mankind, who must act and reason and believe; though they are not able, by their most diligent enquiry, to satisfy themselves concerning the foundation of these operations, or to remove the objections, which may be raised against them" (E 160). For a Pyrrhonian like Hume, there is no metatheory uniting these two dispositions – we move between them carelessly according to the particular situation in which we find ourselves, either in a closet, or in the company of others. After some time in the closet we are pulled back to life and natural reasoning: "Nature, by an absolute and uncontroulable necessity has determin'd us to judge as well as to breathe and feel; nor can we any more forbear viewing certain objects in a stronger and fuller light, upon account of their customary connexion with a present impression, than we can hinder ourselves from thinking as long as we are awake, or seeing the surrounding bodies, when we turn our eyes towards them in broad sunshine" (T 183). This contradiction between reason and instinct is cured by *momentary inattention*, by alternately exposing and hiding one side of the contradiction. The Humean life, claims Popkin, "becomes a continuous alteration between intellectual examination, leading to deeper and deeper abysses the more once tries to understand anything, and periods of relief occasioned by nature's benevolent guidance" (Popkin 1993, 157–8).

The problem is thus not solved on any rational, argumentative grounds but "carelessness and inattention can alone afford us remedy" (T 218). On one hand we accept that we have "no choice left but betwixt a false reason and none at all" (T 268) and on the other we go on with reasoning and judging based on our natural cognitive dispositions, deepening our understanding of the world: "since reason is incapable of dispelling these clouds, nature herself suffices to do that purpose, and cures me of this philosophical melancholy and delirium" (T 269). *This solution does not involve any mitigation of theoretical epistemological scepticism*. Mitigation entails, according to Popkin's provocative argumentation, just the separation of the two powers, reason and instinct; excessiveness lies in the inappropriate attempts of scepticism at extending its jurisdiction; these attempts are, however, only hypothetical – nature would immediately disable them. Why, then, does Hume propose some moderate, mitigated scepticism as an antidote to the Pyrrhonians' excessive scepticism? Does his proposal involve more than what Popkin suggests?

4.4 Hume's Mitigated Scepticism

There are two characteristics that Hume explicitly attributes to mitigated scepticism; both unfold from the awareness of the weakness of our reason. This awareness should firstly tame our pride and "naturally inspire them [men] with more modesty and reserve [that would] diminish their fond opinion of themselves" (E 161). Secondly, it should lead to the "limitation of our enquiries to such subjects that are best adapted to the narrow capacity of human understanding" (E 162). A similar description can be found in the conclusion of the *Treatise*, although Hume does not use the term 'mitigated scepticism'. He recommends diffidence as a guard against any dogmatism that claims to present the only certain or true knowledge; Hume ascribes such claims to all speculative abstruse theories based on superstition of every kind and denomination (T 271). In the *Treatise* Hume does not limit the field of enquiry of true philosophy as strictly as in the *Enquiry*, but the result is roughly the same. True sceptical philosophy should not extend beyond our natural cognitive inclinations; so far, Hume's views are consistent with Popkin's. But soon a new motive emerges: scepticism should not take us to disagreeable subjects. It thus excludes profound sceptical scrutiny concerning the foundations of knowledge on the grounds that such scrutiny is extremely depressing. This negative impact of scepticism *on our emotions* becomes increasingly important for Hume.

The first characteristic, concerning the elimination of dogmatism, is typical of all types of Pyrrhonism.[8] In the extreme it demands that we deny any epistemic value to any proposition; in a more realistic form it recommends curbing our natural enthusiasm and vanity, adopting a more humble, critical attitude as an antidote to "rash arrogance [and] lofty pretentions" (E 41). It aims at cultivating a certain *subjective attitude that has nothing to do with mitigating epistemological scepticism.* The second characteristic of mitigated scepticism shows an unexpected twist in Hume's position. He *prescribes* that we confine our thinking to such subjects that fall under daily practice and experience. Hume demands that mitigated sceptics "will never be tempted to go beyond common life, so long as they consider the imperfection of those faculties which they employ, their narrow reach, and their inaccurate operations" (E 162; similar statements E 41, 150). If taken seriously, this requirement would eliminate profound reflection from philosophy; but even more importantly, it is dogmatic. The ancient Pyrrhonians would never suggest that anything should be spared from scepticism; they would eagerly take apart *any claims,*

[8] Dogmatism was the primary target of the ancient sceptics; apart from the attacks of the well-known sceptics, Timon apparently composed three books of *silli* (lampoons) "in which, from his point of view as a Sceptic he abuses everyone and lampoons the dogmatic philosophers, using the form of parody" (*DL* XII, 111, 521). Montaigne describes the Pyrrhonian response to dogmatists: "can there be any proposition capable of acceptance which it is not right to consider ambiguous?" (Montaigne 1993, 71). Bayle displays desperation when faced with the Pyrrhonian destruction of all knowledge; Pyrrhonists "have a kind of weapon that they call *diallelos* [circular reasoning, infinite regress] ... it is a labyrinth in which the thread of Ariadne cannot be of any help" (Bayle 1991, 423).

from basic observations to philosophical systems. As Montaigne notes, "they want you to contradict them in order to achieve their end: doubt and suspense of judgment"; this may concern the dilemma of whether snow is white or black or whether Aristotle or Plato are right about the eternity of the soul (Montaigne 1993, 70).

For Hume, nature pulls man out of the sceptical pit: "the most trivial event in life will put to flight all his [sceptic's] doubts and scruples" (E 160). The remedy consists in "some avocation, and lively impressions of my senses, which obliterate all these chimeras (T 269); we then carry on with cognizing, philosophizing" – and life. Nature does not provide any arguments that might disqualify scepticism but only strong beliefs that operate on a different level. Why, then, should Hume be concerned about the mitigation of epistemological scepticism – or, rather, why should he view it as excessive? There is no reason for it since he divided the realms of sceptical arguments and positive beliefs. As suggested above, Hume might have been motivated to renounce excessive scepticism *for the effect it has on our frame of mind*. True, we suffer from the negative effects of sceptical reflection only for a while, until rescued by nature, but why should we get into this unenviable mental state in the first place? If we adopt mitigated scepticism and confine ourselves to the observation of human nature, *we avoid these agonizing episodes altogether*. In that case, though, we have to relinquish the capacity for profound reasoning.

This tendency is especially noticeable in the conclusion of Book I of the *Treatise* wherein Hume takes us on the painful journey of a soul to inner peace, achieved in moderate scepticism. Hume opens the journey by pausing for a moment to "ponder that voyage, which I have undertaken, and which undoubtedly requires the utmost art and industry to be brought to a happy conclusion" (T 263). It certainly is a formidable task, considering that the preceding inquiry made him a shipwreck. With all the emotional urgency of youth, Hume presents the deplorable condition in which we find ourselves when indulging in sceptical reflections. This scepticism plunges us into despair (or even illness), as vividly described by Hume in his *Letter to a Physician*, written before his departure to La Fleche (see Appendix 2). On the one hand he observes that though our ideas proceed merely "from an illusion of the imagination … the deficiency of our ideas is not, indeed, perceiv'd in common life"; yet on the other hand, if our awareness of this fact makes us "adhere to … the general and more establish'd properties of the imagination … it wou'd be … attended with the most fatal consequences" (T 267). He has to face the incompatibility of the verdicts of reason and belief which make it almost impossible to answer the question, "what party shall we choose among these difficulties?" (T 268). This is a typical no-win Pyrrhonian trap. Here, Hume reaches the peak of despair, expressed as follows:

> The *intense* view of these manifold contradictions and imperfections in human reason has so wrought upon me, and heated my brain, that I am ready to reject all belief and reasoning, and can look upon no opinion even as more probable or likely than another. Where am I, or what? From what causes do I derive my existence, and to what condition shall I return? Whose favour shall I court, and whose anger must I dread? What beings surround me? and on whom have I any influence, or who have any influence on me? I am confounded with all these questions, and begin to fancy myself in the most deplorable condition imaginable,

inviron'd with the deepest darkness, and utterly depriv'd of the use of every member and faculty (T 269).

It seems that even occasional excursions into sceptical reflections are too traumatic and Hume eventually wonders for what reason and under what obligation must one force oneself to undergo such suffering. Hume reaches breaking point, at which he is ready to exchange this kind of philosophy for some activity that can bring him joy: "I dine", he says, "I play a game of back-gammon, I converse, I am merry with friends" (T 269). Hume acknowledges that sceptical philosophy has nothing to offer to alleviate suffering; but mitigated scepticism can save us from an overdose of the splenetic humour induced by radical scepticism. It can improve our temper *by taking us away from the study*. This rejection of Pyrrhonian excessive scepticism is ironically done in order to embrace another Pyrrhonian goal according to which "the causal principle of scepticism we say is the hope of becoming tranquil" (*PH* I, 12, 5). Instead of indulging in permanent doubts that destroy all propositions (as did the ancient Pyrrhonians) we are told to avoid sceptical scrutiny in those areas for which it is inappropriate. In Hume's specific application of Pyrrhonism its various features collide: to highlight the importance of one feature the other gets suppressed. In this case, *in order to feel agreeable* we should focus on issues that are beyond scepticism's jurisdiction and in which mitigated scepticism can be applied. Mitigated scepticism thus reflects Hume's concerns for common life and anticipates his turn to moral philosophy in which profound scepticism has no place. This intention is supported by Hume's emphasis on the emotional agony that is caused by profound scepticism, and also by his considerations that profound scepticism brings no benefit to mankind.

Hume's introduction of mitigated scepticism in the *Enquiry* is more abrupt. It loses the emotional urgency of the *Treatise* in which, besieged by these Pyrrhonian spectres, Hume "dreads the storm which beats upon me from every side" (T 264). In the *Enquiry* Hume adopts a more pragmatic stance and concentrates on the benefits of true philosophy. At the beginning he mentions the pensive melancholy introduced by abstruse philosophy that is unsuitable for human science – such that has a direct reference to action and society (E 9). Profound reflection, leading to unmitigated scepticism, has similar effects to abstruse metaphysical speculations in preventing us from entering into common life. At the end of his epistemological enquiries, Hume acknowledges the dual condition of mankind; he reaffirms his correction of ancient Pyrrhonism, rejecting its claim that scepticism should have any *constant* influence on the mind. Yet, as is the case in the *Treatise*, Hume is not satisfied with this outcome and introduces mitigated scepticism, which entails a *correction* of excessive scepticism.

This claim, if it were addressed to epistemological scepticism, would be *ad hoc*, unsupported by the logic of Hume's argument throughout his writings on understanding. The alternative suggested here is to view mitigated scepticism as a position that has relevance outside epistemology. Mitigated scepticism reflects the fact that men are active, emotional and social beings for whom it is desirable to follow mild sentiments in order to create a civilized world; man cannot involve himself in the study of morality and in common life intermittently haunted by agonizing contradictions. In addition, Hume says that excessive scepticism brings benefit to neither life nor society: "no durable good can ever result from it" (E 159), while mitigated scepticism "may be both durable and useful" (E 161). More support for mitigated scepticism consists in improving our temper; it helps men to "escape from a state, which to them is so uneasy" (E 161); accordingly, it "strikes in with no violent passion on the mind" (E 40). A mitigated sceptic is portrayed as a "just reasoner" who employs "a degree of doubt and caution, and modesty" (E 161–2). None of these arguments in favour of mitigated scepticism are relevant to epistemology.

Even though mitigated scepticism explicitly appeals to such positive principles of our behaviour, Hume imposes alarmingly dogmatic restrictions on how we are allowed to think. Led by good intentions to spare man from mental disorder, he strictly forbids any inquiries which go beyond common life. This demand is enforced quite aggressively, especially in the *Enquiry*. In the *Treatise* Hume, quite desperate, on the verge of a nervous breakdown and contemplating quitting philosophy forever, is willing "to throw all [his] books and papers into the fire" but in the *Enquiry* he makes a calculated decision only to burn those books that lead us beyond the areas of the allowed researches – such permitted fields of enquiry being mathematics and experimental reasoning. Besides, in the *Treatise* Hume only wanted to burn his own books and papers in order to be able to enjoy the pleasures of life; in the *Enquiry* he planned to plunder libraries as a matter of principle. Pyrrho must have been turning in his grave.

The end of Book I of the *Treatise* is less dogmatic in this respect. Hume allows for researches that might go beyond common life; after all, as he says, one gets too hot when exposed to the sunshine for too long, and too tired after a long time in the company of others. However, on closer examination he is not so generous. What kind of inquiries are we permitted to develop in the closet? Obviously, we are encouraged to argue against metaphysics – this is a constant in Hume's philosophy. As far as sceptical investigations are concerned, we should pursue only those to which we are *naturally inclined*. However, we are certainly not inclined to develop any profound reasoning in which reason subverts itself; the agony of it, the paralyzing effect and the absence of any benefits for life rule it out. Therefore, Hume's proclamations that "in all the incidents of life we ought still to preserve our scepticism" and "if we are philosophers, it ought only to be upon sceptical principles" (T 270), must be interpreted with care. Hume describes true sceptical philosophy, practiced by true philosophers, as very different from the radical scepticism applied in epistemological inquiries; *Pyrrhonism in epistemology is not surpassed by true scepticism but bypassed and ignored*, with true (mitigated) scepticism operating in the domains of common life and morality. Hume's position was a result of two strong influences: Pyrrhonism and the Scottish Enlightenment. "Hume is trying to serve two masters, Pierre Bayle in the philosophical closet, and Francis Hutcheson in the normal world where people have beliefs, theories, etc." (Popkin 1993, 274). In this view, Hume's epistemological scepticism is insuperable on the theoretical level, yet *never leaving the study*.

Robert Fogelin (1983, 1985) develops this view and argues that Hume never abandoned unmitigated epistemological (or theoretical) scepticism and that his position did not change between the *Treatise* and the *Enquiry*. Fogelin introduces several types of mitigated scepticism that can be identified in Hume. It is prescriptive mitigated scepticism that prevents us from suspending beliefs, practising mitigated scepticism that prevents us from applying radical scepticism in life and later he adds conceptual mitigated scepticism that ensures the intelligibility of a system of beliefs that arise in daily life and are understood by an ordinary man (unmitigated conceptual scepticism being limited to philosophical conceits). In sum, prescriptive scepticism is unachievable, practising scepticism that lacks belief would be suicidal and the intelligibility of beliefs is natural since our knowledge is based on natural psychological faculties. Fogelin agrees that Hume's epistemological scepticism is no moderate, Academic, probabilistic skepticism, and offers no middle way between naïve acceptance and Pyrrhonism. The area in which the mitigated types of scepticism operate cannot be justified from the radically sceptical perspective: "there can be no arguments justifying a more mitigated version of scepticism … we find ourselves there" (Fogelin 1985, 150). However, mitigated scepticism is for Fogelin a result of a clash between doubts and instincts in epistemology; I view it as an approach emerging from our being deeply rooted in common life and morality and applicable to this field.

Fogelin advances the thesis that Hume's epistemological, unmitigated scepticism coheres with his naturalistic program and both are mutually supportive. They may be supportive in the sense that we are not purely doubting but also living creatures with animal instincts, imagination, fantasy, etc. Therefore, Fogelin suggests that Hume describes a variety of cognitive perspectives – summed up as that of a country gentlemen (vulgar), of a despairing scepticism (Pyrrhonian) and of the wise (true philosophy) (Fogelin 2009). Men of letters, however, are not spared Pyrrhonian scepticism or the vulgar assent to beliefs. Fogelin's perspectivism does not change the fact that men of letters are *both theoretical sceptics and natural dogmatists* as far as epistemology is concerned. Men of letters may realize that our spontaneous (vulgar) consent to beliefs has no rational basis but their beliefs are not weaker as a result nor can they escape the 'delirium' *if they pursue theoretical scepticism with regard to knowledge*. Men informed of the hopelessness of this scepticism gain one advantage, however – they can accept it in the Pyrrhonian fashion and turn instead to other, more constructive areas of inquiry.

Don Garrett (2004) argues against Fogelin that mitigated scepticism is the culmination of Hume's naturalistic intentions in epistemology.[9] Garrett proposes an even more complex classification of the various types of Hume's scepticism and makes cross-comparisons between them with the main aim of demonstrating that beliefs can have a rational basis.[10] In Garrett's view, Hume is not an unmitigated theoretical

[9] I choose Garret among the many who dispute this 'unmitigated' view because he directly argues against Fogelin, and is also a proponent of the now widely-supported naturalistic interpretation of Hume.

[10] For this purpose I wish to mention the following classification: "we can and should distinguish among three different properties of beliefs: production by reason, epistemic merit, and rational support. A belief is produced by reason if and only if it results from an operation of the inferential faculty. A belief has epistemic merit if and only if it deserves or is worthy of belief or assent. … Finally, a belief has rational support if and only if it has epistemic merit because of the manner in

sceptic "for he is neither a general unmitigated rational support skeptic nor a general unmitigated epistemic merit skeptic" (Garrett 2004, 89). Hume begins with a naturalistic account on how beliefs are formed; that is linked to the negative phase in which we intensely contemplate the lack of rational support for beliefs. However, as Garrett claims, "the 'negative' phase of his [Hume's] standard strategy, in which he shows that certain beliefs are not produced by reason, is not itself skeptical, but naturalistic" (Garrett 2004, 89–90). Any negative, unmitigated sceptical conclusions should be deferred until the end passages of Books I of the *Treatise* and the *Enquiry* wherein Hume asks the crucial question of whether his findings on our mental faculties allow us *to approve of their continued application.*

Garrett focuses on several passages at the end of Books I of both Hume's *Treatise* and his *Enquiry* that seem to be at odds with the more sceptical reading supported herein. He bases his interpretation on the Title Principle, comprising of two brief sentences from Hume's *Treatise*: "Where reason is lively, and mixes itself with some propensity, it ought to be assented to. Where it does not, it never can have any title to operate upon us." (T 270). Garrett claims that in this passage Hume rejects refined and elaborate reasoning because it is "strained", too remote from our inclinations and interests. The new reason, mixed with other propensities such as curiosity and ambition, has, in Garrett's view, almost a miraculously reconciling function. It enables us to reject flights of fancy but to accept permanent, irresistible and universal beliefs; it can allegedly be adopted without self-contradiction because it *already arises from our natural needs*, best suited to satisfy those desires we naturally have. Therefore, our acceptance of those beliefs *has rational grounds*. Strict insistence on epistemological contradictions is, according to Garrett, self-defeating: "if human reason judges itself to be imperfect, then reason itself tells us that we must discount to some extent the very scepticism to which it leads us" (Garrett 1997, 236). Reason thus becomes a natural activity, "one that leads us to approve most of its own [natural] operations when we reflect on them in the light of our desires and felt needs" (Garrett 1997, 241). Garrett's main aim is to confirm that certain beliefs can be rationally assented to and that knowledge can have scientific status: "Hume tells us that some beliefs are not just permitted but ought to be assented to, and assented to because they result from reason as employed under certain specified circumstances. But to say that we ought to assent to some judgments because they are deliverances of reason seems to entail that belief in these propositions is rationally warranted. Thus, Hume's concluding adoption of the Title Principle seems incompatible with his maintaining an unmitigated theoretical skepticism" (Garrett 2004, 76).

It is difficult to dispute claims that unmitigated scepticism is self-falsifying because the admission of the incompetence of reason disqualifies rational (logically consistent) arguments that assert this incompetence of reason. There is obviously a logical incoherence entailed in any positive assertion that a Pyrrhonian sceptic can

which it is produced by reason. Hence, a belief might be produced by reason and have epistemic merit, and yet still lack rational support—because its epistemic merit did not derive from its production by reason" (Garrett 2004, 80).

possibly make. This also applies to the reverse situation – if the sceptics are determined to make no assertions they cannot forbid the making of assertions. As Sextus says, he is "determined to determine nothing, not even 'I determine nothing' itself" (*PH* I, 197, 49). Barnes shows that such reflections are ancient and "start from truism: if Sextus is a sceptic, then he does not hold that he is a sceptic" (Barnes 2000, xxii). However, this should not prevent a sceptic from expressing his own opinions and informing us of them; it must, however, be done *without dogmatic insistence on them in the face of opposing views and without insisting on their objective universal validity*: "although he may not believe anything, Sextus may yet pontificate; although his mind may be empty, his mouth may gush" (Barnes 2000, xxii). It is up to others to undermine the arguments. Hume himself ends Book I of the *Treatise* in this tone; on many an occasion we forget our modesty and we insist that this or that is evident or certain or undeniable; but, he says, "I here enter a *caveat* against any objections, which may be offer'd on that head; and declare that such expressions were extorted from me by the present view of the object, and imply no dogmatical spirit, nor conceited idea of my own judgment" (T 274).

Besides, naturalistic interpretations do not appreciate the historical context of Hume's philosophy. Hume lived in a different time, when (a) epistemological questions were of primary importance, (b) epistemology was a normative discipline and (c) the rational grounding of knowledge was required. Knowledge was not justifiable by the mind's own natural operations. Hume set out to study the nature of ideas – on the one hand, a philosopher must proceed as a geographer, observing and delineating distinct parts and powers of the mind; on the other hand, though, these powers are judged against normative, previously defined rules for the conduct of the mind; in Hume's case that is the copy principle. The natural inclinations of the mind cannot be justified by themselves, by their power or productivity or success. The copy principle ensures that knowledge so achieved is sound and has the credentials of reason. Once it is revealed that our mental operations break the copy principle there is no way around the fact that these credentials are lacking. For Hume, knowledge is grounded in the belief that belongs to our animal nature; beliefs can be described as 'reasonable' only in this sense. Reason as conceived by early modern philosophers was a privilege, a source of norms and the ultimate power placing man above nature; it had the authority to define general methodological rules for science. That is why the modern sceptics were so worried about the weakness of reason and understood the damaging effect of scepticism on the status of knowledge. Naturalist tendencies could be found rather in the Scottish school of common sense, by which Hume had been inspired, but he was a sophisticated philosopher, one who could not dismiss the sceptical assault on the legitimacy of common sense. Naturalism would thus better fit Reid than Hume.[11]

[11] Garrett's account fits better to Reid than to Hume; as Reid says: "The sceptic asks me, Why do you believe the existence of the external object which you perceive? This belief, sir, is none of my manufacture; it came from the mint of Nature; it bears her image and superscription; and, if it is not right, the fault is not mine: I even took it upon trust, and without suspicion. Reason, says the sceptic, is the only judge of truth, and you ought to throw off every opinion and every belief that is

Of course, from the viewpoint of contemporary philosophy, naturalism could be a solution to Hume's scepticism – but it was not then available to him. If Hume had lived today and had been familiar with cognitive science and neurobiology, genetics, DNA and the theory of evolution he may have argued that reason should be understood as a natural capacity along with instincts, beliefs and imagination, all belonging to human nature, which had been evolving for a long time; beliefs could then have been rationally justified or, rather, have justified themselves by the mere fact that they were the result of natural evolution and appropriate to our needs. However, then Hume would have formulated neither his copy principle nor his associationist account of knowledge production – and Hume would not have been Hume.

4.5 Mitigated Scepticism for Life

Mitigated scepticism signals Hume's *hasty retreat from the epistemological field.* This retreat was unsystematic but Hume was impatient to focus on common life and to move on to moral philosophy. In this area, the depressing feelings resulting from excessive scepticism are kept away and man can maintain a balanced state of mind. This balance is acquired if we keep the Pyrrhonian ideal of modesty but also a degree of doubt – a small tincture of Pyrrhonism in Hume's formulation – recommended to "abate the pride of the learned and remind them of the universal perplexity and confusion, which is inherent in human nature" (E 161). Garrett uses this statement to prove that Hume mitigated the force of epistemological scepticism. I suggest that it should be viewed as an attitude suitable for common life, not mitigating or replacing theoretical scepticism. Similar view is advanced, for instance, by Paul Russell who claims that Hume is *both* a Pyrrhonist and a mitigated sceptic: "in the philosophical [theoretical] sphere, insulated from the demands of the ordinary occupations and concerns of common life, Pyrrhonism is irrefutable" (Russell 2008, 209); in common life, however, these principles are unliveable. Hume's 'just reasoner' is thus not a transformed (improved) Pyrrhonist but someone who realizes that Pyrrhonian scepticism has limited use.

Annette Baier, in her impressive book (1991), argues that Hume's mitigated – or true – scepticism develops only *after Hume's epistemological investigations* and interprets the conclusion of Book I of the *Treatise* accordingly; the explosion of Hume's despair that overheats the brain indicates that "we are already making the crucial Humean turn, from intellect to feeling" (1991, 20). According to Baier, Hume's scepticism can be properly understood in the context of Books II and III; the unity of all the books of the *Treatise* enables us to understand Hume's main

---

not grounded on reason. Why, sir, [says Reid] should I believe the faculty of reason more than that of perception?—they came both out of the same shop, and were made by the same artist; and if he puts one piece of false ware into my hands, what should hinder him from putting another?" (Reid 1823, 200–1). The 'shop' or the "artist', if understood as nature (and not God), provides an excellent naturalistic picture of knowledge.

philosophical intention, namely to make reason social and passionate. Hume's encouragements to carry on with sceptical researches, after the company of friends gets tiresome, indicate already a *different form of scepticism*, integrated into common, moral and social life: "the philosophical enterprise in this new phase is and knows itself to be an indulgence of sentiments, a relief from uneasiness" that prepares us for the explorations of Books II and III (1991, 22). Reason appears in a new guise, teamed up with calm passions: "it is accompanied by other abilities and virtues; it is answerable to the shared moral sentiments, it is itself a social capacity … it includes habits and customs … is moralized and made sociable" (Baier 1991, 280–285). It is reason in the service of humanity, contributing to the instruction of mankind.

Hume's focus on this role of philosophy was certainly inspired by Cicero; not by his sceptical views on knowledge as outlined in the *Academica* but by his concern for justice and society, by *tying philosophy to public matters*. This may be the reason why Hume explicitly mentions Academic philosophy and gives it preference over excessive (epistemological) scepticism. It is as if Hume echoed Cicero's words in *De officiis* (written in the form of letters to his son) in which he says, "The principal thing done, therefore, by those devotees of the pursuits of learning and science is to apply their own practical wisdom and insight to the service of humanity … Hence it follows that the claims of human society and the bonds that unite men together take precedence of the pursuit of speculative knowledge" (Cicero 1913, *Off* 1:156, 1: 157). Hume draws on this legacy, claiming explicitly that true philosophy is subservient to the interests of society: "the genius of philosophy, if carefully cultivated by several, must gradually diffuse itself throughout the whole society, and bestow a similar correctness on every art and calling" (E 10). To do that properly philosophy must adopt the ideal of moderation in every way; in scepticism concerning knowledge, in a debate, in actions. Academic scepticism recognizes our fallibility but does not let our doubts paralyze our involvement in any aspect of life, be it knowledge or politics. As Cicero says, "we Academicians are not men whose minds wander in uncertainty and never know what principles to adopt. For what sort of mental habit, or rather what sort of life would that be which should dispense with all rules for reasoning or even for living?" (Cicero 1913, *Off* 2:7). The sceptical empiricists, such as Gassendi and Huet, applied Academic mitigated scepticism to knowledge; they assumed both the general ideal of caution and modesty and the specific epistemological principle that probability is the maximum (and also pragmatically sufficient) goal to be achieved. Hume does not settle for such a vapid compromise in epistemology but embraces the positive message of Academic scepticism to care about public affairs.

Baier's view that mitigated scepticism is a prelude to Hume's moral and political philosophy must be appreciated. Baier, however, goes too far in her claim that Hume's *Treatise* is a unified structure in which the true meaning of epistemological scepticism appears only against the background of the following books. Hume's scepticism should be reconsidered (and mitigated) in retrospect since the Pyrrhonian defective reason, introduced by Hume in the first book, had been transformed in the progress of sentiments that unfolded in the other two books of the *Treatise*. At the end of the *Treatise* reason is, according to Baier, accompanied by other abilities and virtues ["mixes itself with some propensity", as Garrett emphasizes] and, most importantly, is itself a social capacity, both in its activities and in the standards of excellence by which they are judged. Baier's position represents the other extreme of the interpretation of mitigated scepticism. The first extreme (Garrett) consisted in the application of mitigated scepticism to epistemology with the aim of correcting excessive scepticism; Baier places mitigated (true) scepticism in the framework of Hume's moral and political philosophy but claims that due to this process of 'maturing', the self-subverting reason of the profound reflection is elevated to a 'wiser' form.

Donald Livingston also adopts a broad interpretative framework for his explanation of Hume's scepticism. He argues that it must be considered in the light of his philosophy of common life and proposes a dialectical picture of Hume's philosophy. Hume apparently did not share the obsession with epistemology that was typical of modern philosophy; it is common life that "is the governing idea of Hume's philosophy and is internal to his reformed notion of rationality" (1998, xvii). Livingston takes seriously the influence of Pyrrhonism on Hume but considers it too negative. According to him, it is the Pyrrhonian rejection of all authorities and customs, combined with the intention to apply scepticism in common life, that lead to philosophical anxiety and delirium – a very problematic assessment of Hume. Not dogmatism but "the philosophical pride of the heroic [excessive] moment of reflection now appears as a vice, as vanity and arrogance" (Livingston 1998, 37). Sceptical reason is led by the principle of autonomy ignoring the pre-reflexive authority of common life; by contrast, true philosophy brings us back to it: "the true philosopher becomes a loyal participant in a social world requiring the deference of others", claims Livingston (1998, 51). It is the return of a penitent since "the true philosopher knows that … pre-reflective practice is prior to the abstract artifacts of principle, rule, or a theory" (Livingston 1998, 43).

Livingston talks about "Pyrrhonian illumination" as the eye-opening moment (and a moment of penance) in which the previously excessively sceptical philosopher discovers "for the first time the magnificent, philosophically unreflective order of common life in opposition to whatever is constituted by autonomous philosophical reflection" (Livingston 1984, 28). The authority of common life does not come from outside but grows organically from its own practices. Here, the final stage of the dialectical process is achieved, unifying and overcoming the vulgar and false (excessively sceptical) phases; the profoundly sceptical philosopher sheds the autonomy principle and becomes both the true sceptic and the true philosopher. Livingston emphasizes the formative role of custom that provides the background for thinking and naturally commands respect. Livingston's interpretation of Hume offers some interesting links to Husserl and his conception of the *Lebenswelt*; for Husserl, Hume revives the discredited motive of common sense operating in the unquestionable givenness of common world. Husserl defines Hume's problem as follows: "How is the *naive obviousness* of the certainty of the world, the certainty in which we live – and, what is more, the certainty of the *everyday* world as well as that of the sophisticated theoretical constructions built upon this everyday world—

to be made comprehensible?" (Husserl 1970, 96). Husserl wants to further revive Hume's motive of the plain, pre-reflexive self-evidence of our certainty of the world and he does it via a deliberate analysis of consciousness, in which the world provides the only valid, natural horizon its activities. Unlike Livingston, Husserl does not claim that any of this was Hume's own philosophical vision; as a genuine interpretation of Hume it seems overstretched.

Livingston appears to exaggerate the role of custom as the foundational authority of thought. Obviously, the role of customs and tradition was of great importance for both the ancient Pyrrhonians and Hume; it can even be considered in general terms as a determining factor in the constitution of a discourse. Here, Kuhn's concept of a paradigm, or some Foucauldian themes of discursive practices, could be used – for instance, our thought can be considered as being *to an extent* [that varies in different accounts] organized and defined through discourses and their inherent norms and expectations; such a paradigmatic framework includes ways of constituting knowledge, social practices, forms of subjectivity and the power relations between them. From this perspective, another problem would come to the forefront – the scope of our individual freedom of thought within certain all-encompassing 'structures'.

However, this perspective is far from Hume's. For him, backing out of autonomous epistemological scepticism was something of an emergency solution to the sceptical findings concerning our reason – customs safely guide us once reason fails the task; they contain norms and strategies that have proven successful over time and thus deserve respect. It does not seem to be a step towards a dialectical unity in which the contradiction between the vulgar and the overly sceptical is overcome. Yet, as argued earlier, respect for customs and the sceptical scrutiny of our understanding need not be mutually exclusive. In epistemology we can adopt the dual attitude, sceptical in the closet and believing once drawn to light, once led by reason and once by nature. Yet Hume increasingly felt the depressing effects of scepticism; his delirium was caused by an *awareness of the irresolvable contradiction caused by the negative voice of reflection that rejects (theoretically) the positive assent to epistemological beliefs*. This is the source mental discomfort and the reason to abandon scepticism. As in Baier's case, I disagree with Livingston's strategy which fails to give Hume's epistemology the independent value it deserves, despite the fact that Hume found moral and political philosophy more rewarding. Hume was not a systematic thinker such as would write Books I of the *Treatise* and the *Enquiry* with such a grand dialectical unity in mind, planning the transformation of radical scepticism to a 'wiser' form.

Hume does argue that a true philosopher informed of the pitfalls of excessive scepticism and the obscurity of metaphysics, returns to the true path by "renewing his appeal to common sense and the natural sentiments of the mind" (E 9). Yet this return does not imply a sophisticated transformation of how the true philosopher – and a mitigated sceptic – perceives the contradiction between belief and reason in epistemology; an informed person is advised to avoid profound reflection since it leads to no positive resolution, and to rely on common sense as would any other person. A true philosopher is *inattentive* and lets himself be carried only to *constructive* areas of inquiry. Livingston seems to elevate man to the idealized status of a Ciceronian aristocrat who is a perfect, almost a divine creature standing above human weaknesses, mastering strong emotions, detached from sceptical thoughts and transforming them in a dialectical synthesis within the framework of customs. Such a man is, as Livingston claims, "curious, humble, philosophically pious, respecting the totality of customs, patriotic, eloquent, tolerant and extensively benevolent", thus exhibiting his greatness of mind, defined as "an impression of reflection attendant upon consciousness of one's own virtue" (Livingston 1998, 35–52). However, Hume is more human; men are fallible, imperfect, they feel joy and despair, enjoy company, have fondness for wine and entertainment, seek happiness and enjoy the small things in life ('le bon David' of the Parisian salons, a frequenter in the Edinburg clubs).

4.6 Conclusion

Hume faced all the challenges inherent in modern epistemology; he understood that from the perspective of the autonomous mind it is impossible to justify the independent existence of things or relations between them. Unlike other sceptical empiricists such as Gassendi, Huet and Locke, he learnt the Cartesian lesson that the source of our impressions cannot be legitimately located in an independent world. Hume exposed the limits of empiricism showing that the basic ideas we use in our judgment transcend experience Hume also consistently developed Pyrrhonian scepticism, emphasizing that we cannot get to know anything beyond appearances. Moreover we are faced with a contradiction between reason and instinct, a contradiction with no rationally consistent solution. Unlike the French sceptical empiricists – and Locke – Hume did not replace true knowledge with probable knowledge and scepticism with fallibility, but held to sceptical arguments till the bitter end. Due to the insight and consistency of his arguments Hume can be described as an *exemplary modern philosopher*.

Scepticism regarding our knowledge brought Hume to the conclusion that science rests on instincts, not on reason. This conclusion *denies the main ideal of modern philosophy* – establishing rational foundations for scientific knowledge. In the context of modern philosophy, Hume made an extraordinary step – he decided to let it be, to *leave the epistemological problem unsolved and to leave us perplexed*. Instead of trying to overcome epistemological scepticism in any way he decided to move on and focus on other, more enjoyable and rewarding topics, those that are closer to life and aid the prosperity of mankind. In a truly Pyrrhonian fashion he accepted that the impossibility of grounding knowledge on reason is an expression of the whimsical condition of mankind. The *nonchalance* with which Hume brushed away the prominent epistemological concerns of the day, the fact that he shrugged them off with a mere pronouncement, "if we believe, that fire warms, or water refreshes, it is only because it costs us too much pains to think otherwise" (T 270), is astonishing, even *scandalous* for a modern philosopher. This attitude makes Hume a perfect Pyrrhonian and renders him a *lone figure within the modern philo-*

sophical mainstream. Even today, it is difficult for some Hume scholars to digest this Pyrrhonian attitude of his and, as described in this chapter, they try to find some unifying interpretations that would either mitigate Hume's epistemological scepticism – and thus make instincts more allied to reason – or integrate Hume's scepticism within the broad totality of his writing – and thus correct unmitigated scepticism in retrospect, from a 'wiser' perspective.

Hume's perspectivism has far-reaching consequences. It challenges the whole *logocentric* drive, present in modern philosophy, which established an alliance between philosophy, mind, reason, knowledge and science and that culminated in the Enlightenment (and flourishes in today's analytical philosophy). Early modern philosophy built on this ideal. Even the sceptical empiricists of French provenance, plus Locke, who accepted some limitations of reason, perceived it as a misfortune; yet this misfortune was not so acutely felt since they compensated for scepticism through active scientific research in mathematics, astronomy, physics and anatomy. Hume challenged this whole ideal by both appealing to us to *accept the wretched state of reason as a predicament with which we have to live* and offering us a solution that consisted in *relaxing this (sceptical) bent of mind and relying on instincts*. Carefree acceptance, taking one's mind off sceptical pessimism by being merry with friends or walking along the river and absorbing the beauty of nature – how utterly incompatible with the modern spirit!

Hume accepted, truly 'philosophically', with Pyrrhonian calmness, the *essential 'untidiness' of the mind*, in which there are two irreconcilable powers, reason and instinct, with man being placed within each in turn. Although they pull him in opposite directions, they co-exist alongside each other without *any higher unifying perspective*. The perspective of the wise man is available to the educated class but does not exempt even them from the being under the influence of the vulgar view. Hume left the epistemological worries behind with a light heart; as he said, "when we see, that we have arrived at the utmost extent of human reason, we sit down contented; tho' we be perfectly satisfied in the main of our ignorance, and perceive that we can give no reason for our most general and most refined principles, beside our experience of their reality; … And as this impossibility of making any farther progress is enough to satisfy the reader, so the writer may derive a more delicate satisfaction from the free confession of his ignorance" (T xviii).

Ultimately, Hume conveys an optimistic message that once we discover the impossibility of rationally grounding knowledge we can focus on the 'anatomical' investigations of our nature which are free of the anguish caused by profound reflection. In a similar way we then observe men's behaviour in common affairs, in the areas of morality and social life. Here, another strong Pyrrhonian motive comes to the fore. Hume is concerned about how men can achieve a balanced, harmonious state of mind. Investigations into how we form beliefs and the whole area of moral and social philosophy are performed devoid of any sceptical threat. These pursuits are not only beneficial for mankind but also enjoyable for the researchers; no perplexities haunt our mind. The final sections of Book I of both the *Treatise* and the *Enquiry* show how anxious Hume was to leave the problems of the sceptical scrutiny of the foundations of knowledge and to steer the attention to the areas in which we can philosophize in a positive, productive manner – to emotions, religion, politics, law, justice, economics and history.

This dimension of mitigated scepticism goes way beyond Popkin's analysis in which mitigated scepticism has relevance to epistemology. Popkin emphasizes that mitigation does not involve any weak "domesticated" kind of scepticism (of the sort that for instance Garrett proposes) but only a separation of the theoretical (excessive) scepticism on the one hand and the practical natural attitudes on the other; these two worlds do not interact and each holds its own ground. The positive attitude includes our natural ways of acquiring knowledge as well as our behaviour in common affairs. But Popkin seems to underestimate the moral and political domains in which mitigated scepticism is applied irrespective of epistemological issues. In these areas we can practise moderation and help to create a new humanistic culture and society. This society accepts the Pyrrhonian respect for customs and tradition since any forced or forceful social changes entail the danger of violence. Mitigation in society has no counterpart in epistemology where scepticism always has a destructive force.

Hume shares the Enlightenment *ethos* that gives reason the role of a guide to progress both in knowledge and in establishing a better social order. However, for Hume this ideal cannot be accomplished by replacing one dogmatic system with another. Hume's model of the Enlightenment is very different from the French version. The French model demands a radical break with the past – new foundations should be built on the debris of the old order. Accordingly, the religious tradition was replaced by the cult of reason that became as dogmatic and oppressive as the practices of the Church. During the French revolution and in its most extreme form, desecrated churches served as temples of reason; each year a '*Fête de la Raison*' took place instead of Christian festivals. This *fête* had a grandiose form, with the goddess of reason – in an outfit resembling the Greek Athena, including the helmet – was lowered down on a pedestal from the ceiling. Hume promoted a different sort of reason in common life: critical, tolerant, and cautious. The enthusiastic and militant version of the Enlightenment contained all that Hume fought against – fanaticism and the lack of respect for tradition. Hume, by contrast, favours a strategy that avoids excesses and is permanently concerned with providing the right circumstances for the development of an enlightened society that promotes mild sentiments and the dignity of virtue, and ensures that the well-being of mankind is not achieved at the cost of its individuals. Hume's vision is, in a way, close to the Popperian ideal of the open society. Hume was a fierce critic of such institutions that oppress freedom of thought and forbid open discussion. Critical, moderately sceptical and socially sensitive reason is, according to Hume, the only faculty that can instruct us in "the tendency of qualities and actions, and point out their beneficial consequences to society and to their possessor" (Hume 2005, 285).

However, no odes on this reconstructed reason can make us mitigate epistemological scepticism. Mitigated scepticism, whether we call it a just form of philosophy, natural reasoning or true scepticism does not affect the destructive force of intense sceptical reflection on our natural cognitive mechanisms. Hume's mitigated scepticism captures our *natural inclinations* the pursuit of which is *agreeable*; these sceptical inquiries do not draw us into contradictions or negatively affect our mood. They can be described a *reverie*, a contemplation or a critical consideration that may even be somewhat remote from common life – but not as much as to draw us to too deep into abstraction. Mitigated scepticism displays itself in *moral and social life* which is the main domain of positive philosophy for Hume; the first pages of the *Enquiry* highlight his determination to provide – in the science of man – a philosophy that is close to human life and the common world, that cultivates affections, makes us *feel* the difference between vice and virtue, excite and regulate our sentiments, bend our hearts to the love of probity and true honour (see E 5–7).

The science of man is supposed to provide a framework in which various aspects of our nature are balanced. The narrow bounds of human understanding are complemented by our sociable nature, inviting men to enjoy the agreeable company of others and to be active in various occupations; "it seems", Hume says, "that nature has pointed out a mixed kind of life as most suitable to human race, and secretly admonished them to allow none of these biasses to *draw* too much, so as to incapacitate them for other occupations and entertainments" (E 9). Hume's main argument against excessive scepticism does not come from the domain of epistemology. His objections are emotional and social; excessive scepticism understood as scepticism overwhelming our lives is fictitious since it has not this power (as Hume corrected Pyrrho); but the trauma experienced during sceptical episodes could persist even afterwards and thus be an obstacle for reaching the balanced happy life.

In contrast, true philosophy, practising mitigated scepticism, has many benefits. It helps us to achieve a state of happiness by freeing us from the obscurity of metaphysical speculation and from the suffering involved in profound reflection. It also "may contribute to the entertainment, instruction, and reformation of mankind" (E 5). However, all these advantages of mitigated (true) scepticism do not affect Hume's wholly unmitigated epistemological scepticism that leads to what can be called a legitimation crisis of reason. Hume acknowledges that this scepticism is insoluble but it is also toothless beyond its own domain of reflection: "its principles [cannot] retain any influence over our conduct and behaviour" (E 7). Careless reasoners thus have no cause for concern; they easily forget about scepticism and embrace the creative potential of our nature in building knowledge and society, and in cultivating our sentiments. Admitting this duality as the constitutive feature of the human constitution is indicative of Hume's Pyrrhonism. The effort to implant mitigated scepticism into epistemology would result in the 'milk-and-water scepticism' robbing Hume's epistemology of its most original and provocative features.

References

Baier, A. 1991. *A progress of sentiments: Reflections on Hume's Treatises*. Harvard: Harvard University Press.

Barnes, J. 1998. The beliefs of a Pyrrhonist. In ed. M. Burnyeat and M. Frede, 58–92.

Barnes, J. 2000. Introduction. In PH xi–xxxi.

Bayle, P. 1991 [1702]. *Historical and critical dictionary*. Selections, Trans. and ed. R. Popkin. Indianapolis/Cambridge: Hackett.

Bayle, P. 2000 [1682]. *Various thoughts on the occasion of a comet*. Trans. Robert Bartlett. Albany: State University of New York Press.
Burnyeat, M. 1998. Can the sceptic live his scepticism? In ed. M. Burnyeat and M. Frede, 25–57.
Burnyeat, M., and M. Frede (eds.). 1998. *The original sceptics: A controversy*. Indianapolis/ Cambridge: Hackett.
Cicero 1913 [44 BC]. *De oficiis*. Trans. W. Miller. Cambridge, MA: Harvard University Press.
Fogelin, R. 1983. The tendency of Hume's scepticism. In ed. M. Burnyeat, 397–412.
Fogelin, R. 1985. *Hume's skepticism in the treatise of human nature*. London: Routledge and Kegan Paul.
Fogelin, R. 2009. Hume's skepticism. In *The Cambridge companion to hume*, ed. D.F. Norton and J. Taylor, 209–238. Cambridge: Cambridge University Press.
Fosl, P.S. 1998. The bibliographic bases of Hume's understanding of Sextus Empiricus and Pyrrhonism. *Journal of the History of Philosophy* 16: 93–109.
Fosl, P.S. 2011. Skepticism and the possibility of nature. In *Pyrrhonism in ancient, modern, and contemporary philosophy*, ed. D. Machuca, 145–171. Dordrecht: Springer.
Frede, M. 1998a. The sceptic's beliefs. In ed. M. Burnyeat and M. Frede, 1–24.
Frede, M. 1998b. The sceptic's two kinds of assent and the question of the possibility of knowledge. In ed. M. Burnyeat and M. Frede, 127–151.
Garrett, D. 1997. *Cognition and commitment in Hume's philosophy*. Oxford: Oxford University Press.
Garrett, D. 2004. A small tincture of Pyrrhonism. Skepticism and naturalism in Hume's science of man. In ed. W. Sinnott-Armstrong, 68–99.
Hankinson, R.J. 1998. *The sceptics*. London/New York: Routledge.
Huet, P.D. 1725 [1723]. *A philosophical treatise concerning the weakness of human understanding*. London: Gysbert Dommer.
Hume, D. 1998. *Dialogues concerning natural religion*, ed. J.C.A. Gaskin. Oxford: Oxford University Press.
Hume, D. 2005 [1748]. *An enquiry concerning the principles of morals*. In ed. L.A. Selby-Bigge. Revised P.H. Nidditch. Third edition. Oxford: Clarendon Press.
Husserl, E. 1970. *The crisis of European science and transcendental phenomenology*. Trans. D. Carr. Evanston: Northwestern University Press.
Livingston, D.W. 1984. *Hume's philosophy of common life*. Chicago/London: University of Chicago Press.
Livingston, D.W. 1998. *Philosophical melancholy and delirium: Hume's pathology of philosophy*. Chicago: University of Chicago Press.
Montaigne, M. 1993 [1580]. *An apology for Raymond Sebond*. Trans. M.A. Screech. London: Penguin.
Reid, T. 1823 [1764]. An inquiry into the human mind on the principles of common sense. London.
Russell, P. 2008. *The riddle of Hume's Treatise. Skepticism, naturalism and irreligion*. Oxford: Oxford University Press.
Sinnott-Armstrong, W. (ed.). 2004. *Pyrrhonian skepticism*. New York: Oxford University Press.

Chapter 5
Epilogue: Hume on the Role(s) of Philosophy

Abstract For Hume, true philosophy protects us from metaphysics and superstitions; it also shifts our attention from the excessive epistemological scepticism to more rewarding subjects concerned with human behavior in company and affairs, with moral attitudes and social involvement. True philosophy becomes a guide to a balanced, happy life (Hume's version of *ataraxia* spiced by earthly pleasures), advocates modesty and humility as opposed to dogmatism and fanaticism, and encourages a respect for tradition and customs. However, several questions arise. Does Hume's rejection of (excessive) scepticism on the grounds that it is painful and fatiguing not lead to a demise of critical thinking? Scepticism as the core of criticism should mercilessly scrutinize any phenomena under investigation – it must be excessive to fulfil this function. And further, does Hume's emphasis on easy philosophy not degrade philosophy as such? True philosophers, Hume says, offer views that are natural and obvious and could thus have occurred without the assistance of philosophy; they do not deliver any higher wisdom. What, then, are they for? This anti-elitist and anti-intellectual thrust of Hume's philosophy is unparalleled among his contemporaries

Keywords False philosophy • True philosophy • Practical philosophy

5.1 The Science of Man

A wide range of tasks for philosophy can be identified that correspond to Hume's grand project of the science of man. The science of man explores all the manifestations of human nature and has as its subject-matter our reasoning, feelings, virtues and vices, our behaviour in the company of others and in society; our involvement in commerce and business, the principles of law and justice, and the forms of government. Whether directed inwards into our thought and emotions or outwards to the common world these all "intimately concern human life" (T xvi).

Hume distinguishes between three main parts of the science of man in the *Treatise*. It is logic (epistemology in today's usage) that studies our reasoning faculties and the nature of our ideas; morals and criticism consider our tastes and sentiments – here, broadly speaking, Hume treats our emotions and their motivational

Z. Parusniková, *David Hume, Sceptic*, SpringerBriefs in Philosophy,
DOI 10.1007/978-3-319-43794-1_5

function in our behaviour; the third part, politics, considers men as united in society, and dependent on each other (T xv, *Abstract* T 646). Hume introduces only two areas of research the *Enquiry*, reserving the second – of morals – for both the study of sentiments and society. Both these classifications reflect the structure of Hume's main philosophical texts; Book I, II and III of the *Treatise* and Books I and II of the *Enquiry*. Hume thus encourages philosophical investigations along three lines corresponding to three main traits of our nature – philosophy should study man as a reasonable, active and sociable being (E 8).

In the *Treatise* Hume presents the science of man as the second pillar of the new science – the first being represented by Newton's discoveries of the workings of nature and universe and the second by Hume's discoveries about human nature. But Hume had even higher aim – he attributed to the first, epistemological part of his science of man a foundational role. All sciences, including mathematics, natural philosophy and natural religion are in Hume's account comprehended in it and depend on it. Thus Hume gives epistemology an exclusive and privileged role, even more vital than Newton's physics. Since epistemology studies our cognitive abilities and the operations of the mind it is supposed to be the *foundation of the sciences of the physical world* (and of any kind of science generally). Early on Hume was optimistic about the consistency and soundness of this project. Hume probably hoped that by stipulating strict empiricist rules for knowledge – formation, summed up in his copy principle, he could provide a scientific, rationally justified account of human understanding.

But soon the sceptical findings about the powers of intellect disturbed the original plan. As Hume clarified that our ideas of necessary connection (causation) and distinct existence (external things) are illegitimate if judged by the standards of empiricism, so our reasoning lost its rational support. The limits of reason thus reveal the limits of natural science instead of reinforcing it. The primary task of epistemology then turns inwards to a careful observation of the "extent and force of human understanding", and to the explanation of the "nature of the ideas we employ, and of the operations we perform in our reasonings" (T xv). In the *Enquiry* Hume gave up on the foundational role of epistemology and reaffirmed its purely 'geographical' role; mental geography is the delineation of the distinct functions and powers of the mind, and epistemology is supposed to "separate them from each other, to class them under their proper heads and to correct all that seeming disorder, in which they lie involved, when made the object of reflection and inquiry" (E 13).

As discussed in Sect. 2.2, Hume applies Newton's experimental method to the study of human mental world in his epistemology, and to the study of the human involvement in common world in his moral science. Though in moral science we cannot achieve the same degree of precision and exactness as in natural science we should proceed in the most similar manner possible. In terms of our understanding Hume investigates the creative powers of the mind. The role of epistemology, then, is to define how our ideas are derived from our impressions and how they are connected together by associations and are further reinforced by custom and imagination. Hume calls the unifying force of associations an *attraction* that can be considered analogous to the force of gravitation in Newtonian physics. He downplays the role of associations in the *Enquiry* but retains the psychological account of the mind and analyses how, after repeated experience, we get into the habit of forming the belief in the necessity of the experienced succession or union of impressions.

We can examine how the mechanisms of the mind work but, like Newton, Hume discards all questions about the ultimate causes of these operations. On experimental grounds Descartes' conception of innate ideas is rejected along with all other speculative accounts on how our understanding proceeds; Hume says we must stick to what can be observed. When Hume discusses the idea of necessary connection he notes that after a careful examination we can only experience two events conjoined and not connected; yet we believe in their necessary connection as the result of a habit that arises after we observe the conjunction of two events repeatedly; this habit stimulates our imagination and produces a strong, involuntary belief in the causal connection. What is the ultimate cause of these inclinations? For Hume, such questions go beyond our experience and cannot be dealt with. In this respect, "all we know is our profound ignorance" (E 73); experimental science cannot appeal to any ontological sources of causation – our ideas are derived from our impressions whose ultimate source is unknown. Likewise, the emergence of our ideas or beliefs cannot be ascribed to any supernatural causes, for instance to a volition of an omnipotent divine Maker. Such metaphysical grounding of our mind was alien to Hume; for him, the spiritual dimension of the soul – being constituted by the relation to the divine and partaking in a universal teleology – was an antiquated concept of religious doctrines (yet still alive in Berkeley's epistemology and in Bayle's fideism).

5.2 True Philosophy

Hume's science of man provides philosophy with a diversity of themes but *his method imposes strict limitations on the roles that true philosophy can play*. They become apparent in a confrontation with *what true philosophy cannot be* – starting with metaphysics and superstition of any kind. Hume presses this point in the introductory sections of both the *Treatise* and the *Enquiry* and throughout his writings; metaphysics is a burden of the past unsuitable for the enlightened age in which man is free – his own master. Metaphysics takes man outside "the proper province of human reason" (E 12) and causes error and confusion. While epistemology the errors are laughable, in religion, Hume says, they are dangerous. Hume experienced the oppressive practices of the Church in society, the indoctrination of people based on ungrounded speculations; these dogmatic doctrines spread fear in our minds but can also excite feelings of enthusiasm and lead to fanatical behaviour in which vanity will prevail over modesty and moderation.

As shown in Chap. 1, the temptations of metaphysics "lie in wait to break upon every unguarded avenue of the mind" (E 11). Modesty must be cultivated first in epistemology which reveals how imperfect our intellectual powers are. Scepticism about understanding curbs our inappropriate ambitions – to go beyond experience – and can be in the service of humanity if it is applied in common life. "Nothing is more requisite for a true philosopher", Hume says, "than to restrain the intemperate desire of searching into causes, and having establish'd any doctrine upon a sufficient number of experiments, rest contented with that, when he sees a farther examination would lead him into obscure and uncertain speculations" (T 13). Therefore, true philosophy avoids "all distant and high enquiries, confines itself to common life, and to such subjects as fall under daily practice and experience" (E 161). Morality and politics can thus be set free of the pretentions of the doctrinaires and of the vanity of the dogmatists, and can instead endorse the values of modesty, tolerance and mutual respect instead.

Yet, at the end of the *Treatise* Hume raises an important question. Is it not the case that metaphysical themes attract curiosity in men, and that we are *naturally inclined* to pursue inquiries concerning the deepest principles of our nature, moral values and the foundations of society – in other words inquiries that go beyond the spheres of common sense and common life? After all, "'tis almost impossible for the mind of man to rest, like those of beasts, in that narrow circle of objects, which are the subject of daily conversation and action" (T 271). As Hume observes, people may hope to fulfil their ambitions to contribute to the well-being of mankind by discovering *more profound knowledge*; they consider such knowledge more insightful than the down-to earth observations and they ignore the fact that such inquiries are beyond the faculties of our reason. This is indeed an intriguing dilemma. On the one hand, Hume imposes the methodological rule that we remain within the limits of experience, but on the other hand he considers that we have a natural inclination to go beyond it. Does the observation of human nature not point to a *natural need for metaphysical speculations* (and, perhaps, religious faith)? We could face another sceptical contradiction here. How could Hume dismiss metaphysics from philosophy without being dogmatic or straightforwardly repressive?

At this stage, Hume introduces other two criteria for true philosophy that carry us beyond theoretical epistemology. First, philosophy must be of *practical benefit to society* – on this ground, Hume discards metaphysics that in his view leads only obscurity; abstruse philosophers "raise these intangling brambles to cover and protect their weakness" resulting from their vain endeavour to penetrate into subjects entirely inaccessible to human understanding (E 11). For this reason, metaphysics is of no benefit to knowledge or society. Second, true philosophy must be *pleasing* and philosophizing must *feel agreeable*. Metaphysics – and also profound scepticism are dismissed on the grounds of being *painful* and *fatiguing*, and *easy philosophy* is put in their place (E 11). Hume praises easy philosophy for awakening positive emotions and being more useful in life: "it [easy philosophy] enters more into common life; moulds the heart and affections" (E 6–7). Ultimately, Hume dismisses both metaphysics and excessive scepticism for pragmatic and psychological reasons.[1]

[1] There is a curious twist in Hume's position in his essay "Of Commerce"; in the introductory passage he distinguishes between shallow and abstruse thinkers and defends the latter. He appreciates the ability of the abstruse thinkers (metaphysicians and refiners) to produce fine discoveries, to say something uncommon and new and to regard the general course of things. In contrast to his key

The latter criterion of true philosophy shows Hume's affiliation with Pyrrhonism; Hume understood that the concern of the Pyrrhonians was not just epistemological but entailed the search for the ways of coping with scepticism *in order to put the mind at rest*, to escape mental disturbances and suffering. The ancient sceptics chose detachment from life and indifference to disturbance; the original connotations of *ataraxia* were tied to a kind of *apatheia* in life. Hume chose a different way of coping with scepticism – he immunized our natural inclinations and behaviour against its influence. This strategy enabled him to give moral philosophy a more active and optimistic role. Drawing on the message of the Pyrrhonians Hume believed that *philosophy must help us to achieve happiness*. But Hume did not see happiness in detachement but rather placed man in the bustle of life, and added a more frivolous flavour to *ataraxia* by appealing to the elements of pleasure and joy. Accordingly, Hume emphasized the advantages of easy and obvious philosophy over the abstruse kind: "the feelings of our heart, the agitation of our passions, the vehemence of our affections … reduce the profound philosopher to a mere plebeian" (E 7). Hume said repeatedly that true philosophy was closer to the vulgar than to the 'subtile' approach, and he appreciated the "gross earthy mixture" of the country gentlemen who "being always employ'd in their domestic affairs, or amusing themselves in common recreations, have carried their thoughts very little beyond those objects, which are every day expos'd to their senses" (T 272).

It seems that true philosophy has for Hume something of a 'policing' or disciplining role; setting limits to inquiry rather than extending philosophical ambitions. Here, as Hume says, "philosophical decisions are nothing but the reflections of common life, methodized and corrected" (E 162). Methodized reflections provide a methodical, systematic account on various aspects of common life, this claim is uncontroversial. Corrected reflections are free of any speculations that go beyond experience. Hume's essay on miracles can serve as a good example. Men sometimes tend to misinterpret the testimony of the senses; they draw illegitimate conclusions from the observed instances, and as this 'testimony' is transferred from one generation to another it becomes ever more far-fetched. True philosophy thus unmasks the falsity of reports of fantastic or miraculous events. More generally, it guards common life against superstitions and metaphysics that, as Hume points out, we are sometimes naturally drawn into in the false hope of making grand discoveries. However, these are only dangerous illusions; for Hume "any hypothesis, that pretends to discover the ultimate original qualities of human nature, ought at first to be rejected as presumptuous and chimerical" (T xvii).[2]

arguments in the *Treatise* and the *Enquiry* Hume ridicules shallow (easy) thinkers for not being able to go beyond their own weak conceptions and considers abstruse thinkers more valuable. This somersault seems inexplicable unless Hume thinks of both kinds of thinkers in this specific context of common life (of commerce); abstruse thinkers would not be led by flights of fancy beyond the sphere of this life but would be more able to dig out intriguing relations within it, while shallow thinkers would be nearly dumb, immune to any intellectual stimulation.

[2]The arguments in this paragraph were suggested to me by a communication from James Hill.

As analyzed in previous chapters, Hume's epistemological scepticism remains wholly unmitigated and cannot be cured by argument; any attempted inquiries into these matters would become a branch of metaphysics. This conclusion is reached by a reflection that reveals human blindness and weakness. But the force of nature that saves us from its paralyzing effect by exerting its irresistible primordial power over our theoretical thinking; thus it brings us on the right, constructive path of common sense. This duality of human constitution gave Hume a chance to preserve excessive scepticism in the epistemological domain and yet to develop an active account of man's cognitive procedures and achievements. This picture allows for both merciless scepticism and cognitive optimism, provided that we admit that knowledge has no rational justification – and Hume has no qualms about doing that. A philosopher thus can switch between the two modes of existence, the reflective and the instinctive, being a sceptic and a common man by turn. But at the end of Books I of the *Enquiry* and especially of the *Treatise*, Hume seems not too keen on switching to and fro; he revives the Pyrrhonian concern about *the undisturbed state of mind* and concludes that scepticism, though pursued in limited periods in the study, has the opposite effect – it makes one desolate. Why should we undergo even one more episode of profound scepticism? If we get tired of company and want to devote time to philosophical researches, he says, we should limit ourselves to those that are *agreeable*. And that radical scepticism is not. Does this mean that once the sceptical job regarding our understanding is done, there is no need to waste time on this topic again? It seems so. Hume is ready to denounce radical scepticism (in the same way he denounced metaphysics) since it makes us feel bad, and to opt for the Academic (true) scepticism that makes us feel good. Though on the theoretical level Hume remained an unmitigated sceptic, he felt uncomfortable about the psychological effects of unmitigated scepticism.

What, then, is the role of radical scepticism in Hume's philosophy? One clue is given by Hume's usage of terminology. True scepticism is for him the mitigated kind, taking from unmitigated scepticism the memento of the weaknesses of our intellect and thus making us modest in our ambitions. But does this form of scepticism deserve to be called 'scepticism' at all? Scepticism should mercilessly scrutinize any phenomena under investigation – it must be excessive to fulfil this function. Mitigated scepticism is thus a contradiction in terms; it should be rather called a mildly critical or easy approach.[3] There is only one scepticism worth the name 'true' scepticism, which is excessive by definition; scepticism as *a method must*

[3]The concept of mitigated scepticism reminds me of a political parody, written by Jaroslav Hašek (the author of the novel The Good Soldier Švejk), in which he forms "The party of moderate progress within the bounds of the law" (and is finally nominated for a seat in the Austro-Hungarian parliament). The name of the party reflected the political situation of that time in which no criticism of the government was allowed – the name was thus a caricature of the critical approach. Of course, if taken seriously, the name of the party can express the ideal of piecemeal social strategy contrasted with radical calls for revolution; in this respect, Hume and Popper hold hands, and the ideal (entailing criticism on a small scale and in areas that can be painlessly reformed if necessary) can certainly appeal to many. But rational sceptical reflection cannot be diluted to this pragmatic approach in politics (and in Hume's common life).

proceed indiscriminately. Mitigated 'scepticism' by contrast does not require deep (and often painful) thinking and lacks the determination to undermine all knowledge claims. The observation that our cognitive instincts and our behaviour in company are in practice resistant to scepticism, or the observation that without scepticism we feel happier should not be reasons for disabling this unique human faculty.

5.3 False Philosophy

Does Hume describe scepticism as false philosophy? Livingston seems to suggest that in his interpretation of Hume's comparison between the vulgar, false and true philosophy. He argues that sceptical reflection legitimately exposes the inconsistency in vulgar opinions but it goes too far; for Livingston, this reflection is guided by the principle of autonomy that "demands that the philosopher, at least in thought, cease being a participant in the prejudices of common life and imagine himself the … arbiter of them" (Livingston 1998, 18). Well, indeed, one might say – this is the role of scepticism. There always is a certain framework that determines how we live and think, created by customs, culture, discursive practices and background knowledge, but it is the privilege of criticism to breach these limits and to expose as problematic what has been so far taken as self-evident, and as a result to restructure our perception of common life. Scepticism and criticism play the role of an assailant not of a jovial conversationalist. But Livingston views this sceptical approach as mistaken and typical of a false philosopher. His false philosopher is alienated from the world of custom and, ultimately, his acts of philosophical reflection do not cohere with human nature – therefore, Livingston argues, such philosophy cannot provide human self-understanding and is inconsistent with its own nature. It is only true philosophy that completes this dialectical circle, recognizes that it must participate in the world of custom, and even accepts "the primordial authority of custom" (Livingston 1998, 22). The falsity of reflection consists in the fact that "it is not true to what a more coherent philosophical understanding knows to be its real nature" (Livingston 1998, 19). If Hume really held this view he would totally have devalued sceptical reflection as a unique act of critical reason whose highest possible autonomy is not a fault (or dialectical immaturity) but a necessity and a privilege. Why should one eliminate this faculty from human nature?

Hume's own analysis of the three attitudes in the *Treatise* seems less ambitious. The vulgar view is obvious; men rely on common sense and follow custom. Philosophers who reflect upon the foundation of these beliefs (and who abstract from custom) find that the common sense attitude is shaky. However, *Hume does not consider this sceptical reflection false philosophy*. False philosophers do not admit defeat at this stage and continue to search for the qualities in which the idea of power consists; Hume claims that they have sufficient force of genius to expose the errors of the vulgar views "but not sufficient to keep them from ever seeking for this connection in matter, or causes" (T 223). *False philosophers are therefore metaphysicians, not sceptics*. True philosophers, then, return to the vulgar and regard all the previous disquisitions with *indifference*. They are indifferent to both sceptical conclusions and the speculations of metaphysicians. Such a return does not seem to be the grand dialectical unity that Livingston suggests. Of course, there is a difference between the vulgar and the true stance. As an educated man a true philosopher, unlike a country gentleman or any ordinary folk, is *aware* of the irrational nature of vulgar beliefs – but his common sense beliefs are not affected by this knowledge (they are not dialectically reformed). In Hume's account *sceptical reflection thus stands in no-man's land*; neither vulgar nor false nor true. This reflects Hume's ambivalent feelings about the role of scepticism in philosophy.

Hume's rejection of (excessive) scepticism in the *Enquiry* on the grounds that it overheats the brain and brings no durable good can be seen as a *demise of critical thinking*. Hume notes that "sceptical doubt arises naturally from a profound and intense reflection on those subjects [a scrutiny of understanding and the senses]" (T 218). Should we give it up because it strains our mind or exposes contradictions? And how can Hume claim that scepticism does not bring any good? The exposition of contradictions moves our knowledge further and opens new horizons; scepticism challenges us to come forward with new suggestions, and what seems insoluble at present may provoke other attempts at solving later. And even if there does not seem to be any solution – in this case a solution to the lack of rational legitimation of our instinctive cognitive inclinations – still scepticism provides a deep insight into the situation, better self-understanding and stimuli for others to tackle the problem. A turn to metaphysics – poking at the ultimate principles of the mind or the world – may not be the only alternative to epistemological dilemmas.

But there are a few hints to this effect in Hume; he remarks: "'Tis sufficient, if I can make the learned world apprehend, that there is some difficulty in this case, and that who-ever solves the difficulty must say some thing very new and extraordinary" (T 657). However, Hume himself, instead of recommending scepticism as a tool to scrutinize every proposition, hypothesis or system, whether it concerns epistemology or common life, recommends inattention. This situation does not stimulate him, for instance, to rethink the strictness of the copy principle but he closes the issue by arguing that further attempts would only be depressing and useless. As remarked earlier, in social philosophy Hume and Popper can find some overlaps; but Popper's emphasis on the vital role of criticism makes a good contrast to Hume's defeatist approach. According to Popper, we should exploit this *negative* rational capacity since "rational criticism is indeed the means by which we learn, grow our knowledge, and transcend ourselves" (Popper 1988, 27). By *critically* discussing problems and by internalizing criticism (the Humean excessive scepticism) in our thinking we facilitate progress and make knowledge an adventure full of surprises. Hume withdraws from the sceptical battlefield (and from scepticism in epistemology) and his declared – and perhaps admirable – modesty may at the same time appear 'gutless'.

Hume's picture of philosophy without scepticism thus looks somewhat lame. When Hume, once again, rejects any metaphysical deliberations that are full of obscurities and perplexities, he says: "happy, if she [philosophy] be sensible … and leaving a scene so full of obscurities and perplexities, return, with suitable modesty, to her true and proper province, the examination of common life; where she will find difficulties enough to employ her enquiries, without launching into so boundless an ocean of doubt, uncertainty, and contradiction!" (E 103). It must be said, however, that the feeling of helplessness in the face of the sceptical conclusion regarding the weakness of reason was common in that time; Pyrrhonian scepticism was deemed fatal. Philosophers who were affected by Pyrrhonism sought various solutions outside the domain of epistemological arguments; Descartes and Berkeley in metaphysics, Gassendi and Bayle in fideism, and Hume in naturalism.

5.4 Practical Philosophy

Yet what is philosophy without scepticism, eagerly launching itself into doubts in the hope of discovering something new – in other words, what does Hume's true philosophy amounts to? In a nutshell, Hume defines the role of epistemology as mental geography (observation of mental operations), while in common life philosophy "can only observe what is commonly done" (T 268). The role of true philosophy thus seems very pedestrian. In epistemology, Hume established the foundations of the philosophy of mind, a discipline whose boom much later he could not have predicted. Hume himself probably thought that the mental map he drew was definitive; he considered the force of associations in human mind as fundamental as the forces in nature and the universe revealed by Newton. Here, then, the constructive role of epistemology ends. Hume is 'perfectly satisfied' with exposing the limits of our reason and with the conclusion that "we can give no reason for our most general and most refined principles, beside our experience of their reality" (T xviii). Hume's response to the detected contradiction between the mental map and the norms of knowledge-acquisition entailed by empiricism is clear – forget it and follow nature. In sum, "a philosopher, who purposes only to represent the common sense of mankind in more beautiful and more engaging colours, if by accident he falls into error, goes no farther; but renewing his appeal to common sense, and the natural sentiments of the mind, returns into the right path, and secures himself from any dangerous illusions" (E 7).

The dominant role of philosophy, then, lies in the observation of common life, in the inquiries into the principles of morals, including sentiments, passions and public affairs; in these areas, mitigated 'scepticism' should be applied. Are we allowed more than to observe and document what is commonly done? Explicitly Hume does not seem to grant philosophy much bigger role; true philosophy should be easy and obvious, agreeable and entertaining, innocent and undemanding. There is a tendency in Hume's pronouncements to downplay the importance of philosophy. For Hume, philosophy certainly falls from the 'throne' it used to enjoy traditionally, in the era of metaphysics, but it cannot fulfil the typical modern role as being a rational foundation of science either, due to the sceptical findings. Philosophy should thus be anchored in common life and respect its rules; it must not stand above as some norm-giver or fierce critic. In this sense, Hume seems to provide a refreshing picture of philosophers. He shows that true philosophy must have understanding for human needs and feelings, aspirations and weaknesses; its tone is civil, close to ordinary people and their concerns, keeping down to earth – in contrast to the subtile philosophy, aspiring to heights where "the air is too fine breathe in, where it is above the winds and clouds of the atmosphere" (Hume 2008, 107). Hume remarks that a "profound" philosopher who carries out speculations of too great subtlety – a philosopher aiming too high or too deep – is a person "which is commonly but little acceptable to the world, as being supposed to contribute nothing either to the advantage or pleasure of society; while he lives remote from the communication with mankind, and is wrapped up in principles and notions equally remote from their comprehension" (E 8).

True philosophers avoid being overeducated, dry, incomprehensible and outright boring as a result, and, equally, they avoid being too sceptical since that results in an overwhelming depression. Fogelin describes (with a reference to Milton) this ideal of a philosopher as '*lowly-wise*'; his citation from Alexander Pope's *Essay on Man* captures Hume's ideal of a middle (mitigated) way[4] (Fogelin 1983, 410). In contrast to the true (just) philosopher the profound (mere) philosopher contributes nothing either to the advantage or pleasure of society, "while he lives remote from communication with mankind, or is wrapped up in principles and notions equally remote from their comprehension" (E 8). True philosophers are not separated from common life and the company of others. They are *adequately* educated since "the mere ignorant is still more despised" (E 8) but they do not deliver any higher wisdom. True philosopher must achieve the right balance, "retaining an equal ability and taste for books, company and business …. [if so applied] virtue becomes amiable, company instructive, and retirement entertaining" (E 8). But does Hume's anti-elitism not degrade the role of philosophy? Husserl is of this opinion and says: "Hume became the father of the weak positivism that avoids all philosophical abysses" (Husserl 1970, 111). Hume's dismissal of metaphysics, though characteristic for that time, brings the danger of dogmatism in case we find in ourselves a natural tendency to this form of inquiry; the dismissal of radical scepticism from all areas of life – except the profound reflection on our mental operations – reduces philosophers to spectators and informed commentators.

Hume's demeaning statements addressed at philosophy may sound surprising. He contributed a great deal to the theory of morals and to political theory, and his analysis of our emotional life and of the role of passions in our actions is certainly not a result of a 'salon conversation'. Hume formulated a theory of moral sense (drawing on Hutcheson) and virtues (inspired by Shaftesbury), based on the assumption than man is motivated to action by feelings and sentiments; he further developed a detailed account of various kinds of passions and their role in forming our virtues. Hume distinguished between natural virtues (some of which are 'noble', like compassion, generosity or benevolence, and some more personal, like friendship or wit and humour), and artificial, socially acquired virtues like justice or

[4] Placed on his isthmus of a middle state/A being darkly wise, and rudely great:/With too much knowledge for the Sceptic side/With too much weakness for the Stoic's pride.

allegiance. The latter virtues play an important role in political theory, in which another concern comes to the forefront – that of social utility serving the good of the whole society. Hume was faithful to these values when he parted with Pyrrhonism in the context of *life*. On the individual level, philosophy should bring man good feelings and a happy frame of mind; calm passions and virtuous acts are *pleasing*. On the social level, Hume is concerned about the happiness and prosperity of mankind, the achievement of which is possible by our collective involvement and mutual assistance in developing the virtues of benevolence and justice. All these goals are on the agenda of true philosophy.[5]

If these inquiries are characteristic of true philosophy then more is expected of philosophy than an entertaining conversation at a dinner table. When Hume says that "all the philosophy, therefore, in the world … will never be able to carry us beyond the usual course of experience, or give us measures of conduct and behaviour different from those which are furnished by reflections on common life" (E 146), he implies that these reflections are not just observations but also *evaluations* and careful *proposals*, both for personal behaviour and the organization of society. This is the true domain of Academic scepticism in which mild criticism is not separated from feelings and, more generally, from our concerns for the well-being of individuals and nations. Hume, drawing on the legacy of Cicero in linking philosophy to society, considers this role of philosophy in specific areas of social life. He believes that, inspired by the ideals of true philosophy, "the politician will acquire greater foresight and subtility, in the subdividing and balancing of power; the lawyer more method and finer principles in his reasonings; and the general more regularity in his discipline, and more caution in his plans and operations; … the stability of governments will be improving" (E 10). Such philosophical issues may be with all seriousness discussed among men of letters with general education and wide range of interests, even at the dinner table. It is therefore surprising that *Hume himself degrades philosophy*. In his essay "The Sceptic" Hume first considers what philosophy might offer – maybe some particular views that would have otherwise escaped us, some guidance helping us to refine our temper, to be more modest and tolerant? This could grant philosophy a broad humanizing function. But in no time, Hume adds: "if these views be natural and obvious, they could have occurred of themselves without the assistance of philosophy; if they be not natural, they never

[5] In his essay "Idea of a Perfect Commonwealth" he provides a manual how to form a just government. He keeps his conviction that any utopian great reformations of mankind are imaginary and dangerous. Thus Hume, on the most general level, supports decentralization and the separation of powers. But he does not appeal to abstract ideals but proposes specific guidelines for the organization of society concerning, for instance, the number of counties and parishes, the number of representatives, the hierarchy of powers, the formation of a Swiss-like militia, elections and the voting system, the reform of the House of Lords, the restoration of the Cromwell's plan of parliament. He proposes one council for both religion and learning that inspects the universities and clergy. This might seem surprising for the 'enlightened' Hume, given his criticism of the institutional role of the Church – and of its repressive influence over thought (and the careers of Academics) (Hume 2008, 301–315).

can have any influence on the affections" (Hume 2008, 106). In this extreme perspective, philosophy dissolves into common talk.

If we take a less extremist view and acknowledge Hume's positive contribution to all fields of the science of human nature, then we can look at true philosophy more benevolently. In epistemology he was the founder of the philosophy of mind. In moral and political science Hume may be considered a forerunner in establishing (and pursuing) specific disciplines in social philosophy or humanities generally, including history; *The History of England* is his most extensive and systematic work. However, this kind of philosophy transforms itself into social sciences; here Hume encourages us to cultivate a humanistic approach and to get personally involved in all areas of life; on an individual level it brings pleasure and on the community level it brings prosperity to society. This anti-elitist and anti-intellectual thrust of Hume's philosophy is unparalleled among his contemporaries.

References

Fogelin, R.J. 1983. The tendency of Hume's scepticism. In *The skeptical tradition*, ed. M. Burnyeat, 397–412. Berkeley: University of California Press.

Hume, D. 2008 [1741]. *Selected essays*. Oxford: Oxford University Press.

Husserl, E. 1970. In *The crisis of European science and transcendental phenomenology*, ed. D. Carr. Evanston: Northwestern University Press.

Livingston, D.W. 1998. *Philosophical melancholy and delirium: Hume's pathology of philosophy*. Chicago: University of Chicago Press.

Popper, K.R. 1988. *Realism and the aim of science*. London: Hutchinson.

Appendices

Appendix 1

Newton to Boyle

Honoured Sir

I have so long deferred to send you my thoughts about the Physicall qualities we spake of, that did I not esteem my self obliged by promise I think I should be ashamed to send them at all. The truth is my notions about things of this kind are so indigested that I am not well satisfied my self in them, & what I am not satisfied in I can scarce esteem fit to be communicated to others, especially in natural Philosophy where there is no end of fansying. But because I am indebted to you & yesterday met with a friend M[r] Maulyverer, who told me he was going to London & intended to give you the trouble of a visit, I could not forbear to take the opportunity of conveying this to you by him.

It being only an explication of qualities which you desire of me, I shall set down my apprehensions in the form of suppositions as follows. And first I suppose that there is diffused through all places an æthereal substance capable of contraction & dilatation, strongly elastick, & in a word much like air in all respects, but far more subtile.

2 I suppose this æther pervades all gross bodies, but yet so as to stand rarer in their pores then in free spaces, & so much the rarer as their pores are less. And this I suppose (with others) to be the cause why light incident on those bodies is refracted towards the perpendicular; why two well polished metalls cohere in a Receiver exhausted of air: why Quicksilver stands sometimes up to the top of a glass pipe though much higher than 30 inches: & one of the main causes why the parts of all bodies cohere. Also the cause of philtration & of the rising of water in small glass pipes above the surface of the stagnating water they are dipt into: for I suspect the æther may stand rarer not only in the insensible pores of bodies, but even in the very sensible cavities of those pipes. And the same principle may cause Menstruums to

Z. Parusniková, *David Hume, Sceptic*, SpringerBriefs in Philosophy,
DOI 10.1007/978-3-319-43794-1

pervade with violence the pores of the bodies they dissolve, the surrounding æther as well as the Atmosphere pressing them together.

3 I suppose the rarer æther within bodies & the denser without them, not to be terminated in a mathematical superficies but to grow gradually into one another: the external æther beginning to grow rarer, & the internal to grow denser at some little distance from the superficies of the body, & running through all intermediate degrees of density in the intermediate spaces. And this may be the cause why light in Grimaldo's experiment passing by the edge of a knife or other opake body is turned aside & as it were refracted, & by that refraction makes several colours.

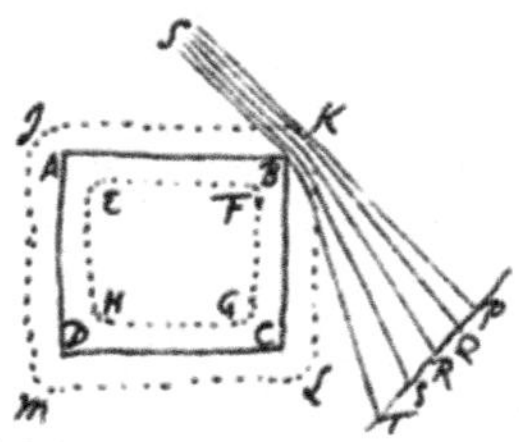

Let ABCD be a dense body whether opake or transparent, EFGH the outside of the uniform æther which is within it, IKLM the inside of the uniform æther which is without it; & conceive the æther which is between EFGH and IKLM to run through all intermediate degrees of density between that of the two uniform æthers on either side. This being supposed, the rays of the sun SB, SK, which pass by the edge of this body between B & K, ought in their passage through the unequally dense æther there, to receive a ply from the denser æther which is on that side towards K, & that the more by how much they pass nearer to the body, & thereby to be scattered through the space PQRST, as by experience they are found to be. Now the space between the limits EFGH & IKLM I shall call the space of the æther's graduated rarity.

4 When two bodies moving towards one another come neare together I suppose the æther between them to grow rarer then before, & the spaces of its graduated rarity to extend further from the superficies of the bodies towards one another, & this by reason that the æther cannot move & play up & down so freely in the strait passage between the bodies as it could before they came so neare together. Thus if

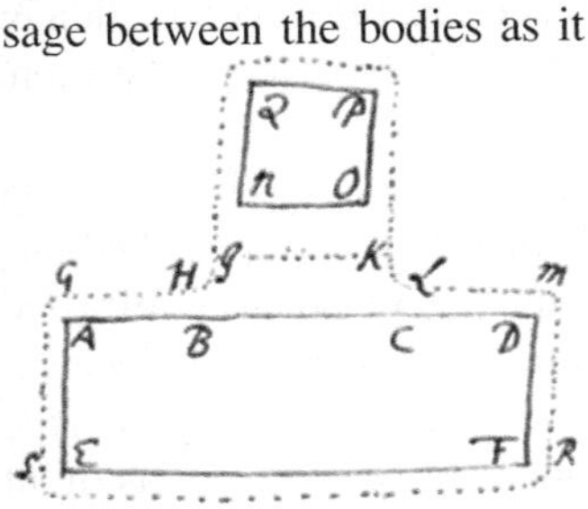

the space of the æther's graduated rarity reach from the body ABCDFE only to the distance GHLMRS when no other body is neare it, yet may it reach farther, as to IK, when another body NOPQ approaches: & as the other body approaches more & more I suppose the æther between them will grow rarer & rarer.

These suppositions I have so described as if I thought the spaces of graduated æther had precise limits, as is exprest at IKLM in the first figure & GMRS in the second: for thus I thought I could better express my self. But really I do not think they have such precise limits but rather decay insensibly, & in so decaying extend to a much greater distance then can easily be beleived or need be supposed.

5 Now from the 4th supposition it follows that when two bodies approaching one another, come so neare together as to make the æther between them begin to rarefy, they will begin to have a reluctance from being brought nearer together, & an endeavour to recede from one another: which reluctance & endeavour will encrease as they come nearer together because thereby they cause the interjacent æther to rarefy more & more. But at length, when they come so neare together that the excess of pressure of the external æther which surrounds the bodies, above that of the rarefied æther which is between them, is so great as to overcome the reluctance which the bodies have from being brought together: then will that excess of pressure drive them with violence together & make them adhere strongly to one another, as was said in the second supposition. For instance in the second Figure when the bodies ED & NP are so neare together, that the spaces of the æthers graduated rarity begin to reach to one another & meet in the line IK; the æther between them will have suffered much rarefaction which rarefaction requires much force that is much pressing of the bodies together: & the endeavour which the æther between them has to return to its former natural state of condensation will cause the bodies to have an endeavour of receding from one another. But on the other hand to counterpoise this endeavour there will not yet be any excess of density of the æther which surrounds the bodies above that of the æther which is between them at the line IK. But if the bodies come nearer together so as to make the æther in the mid-way-line IK grow rarer then the surrounding æther, there will arise from the excess of density of the surrounding æther a compressure of the bodies towards one another: which when by the nearer approach of the bodies it becomes so great as to overcome the afforesaid endeavour the bodies have to recede from one another, they will then go towards one another & adhere together. And on the contrary if any power force them as under to that distance where the endeavour to recede begins to overcome the endeavour to accede, they will again leap from one another. Now hence I conceive it is chiefly that a fly walks on water without wetting her feet, & consequently without touching the water; that two polished pieces of glass are not without pressure brought to contact, no not though the one be plain, the other a little convex; that the particles of dust cannot by pressing be made to cohere, as they would do if they did but fully touch; that the particles of tinging substances & salts dissolved in water do not of their own accord concrete & fall to the bottom, but diffuse themselves all over the liquor, & expand still more if you ad more liquor to them. Also that the particles of vapors exhalations & air do stand at a distance from one another, & endeavour to recede as far from one another as the pressure of the incumbent atmosphere will let them: for I conceive the confused mass of vapors air & exhalations which we call the Atmosphere to be nothing els but the particles of all sorts of bodies of which the earth consists, separated from one another & kept at a distance by the said principle.

From these principles the actions of Menstruums upon bodies may be thus explained. Suppose any tinging body as Cochineel or Logwood be put into water, so soon as the water sinks into its pores & wets on all sides any particle, which adheres to the body only by the principle in second supposition: it takes of or at least much diminishes the efficacy of that principle to hold the particle to the body because it makes the æther on all sides the particle to be of a more uniform density then before. And then the particle being shaken of by any little motion, flotes in the water, & with many such others makes a tincture; which tincture will be of some lively colour if the particles be all of the same size & density, otherwise of a dirty one. For the colours of all natural bodies whatever seem to depend on nothing but the various sizes & densities of their particles: as I think you have seen described by me more at large in another paper. If the particles be very small (as are those of salts Vitriols & gumms) they are transparent, & as they are supposed bigger & bigger they put on these colours in order black, white, yellow, red; violet, blew, pale green, yellow, orange, red; purple, blew, green, yellow, orange, red &c: as is discerned by the colours which appear at the several thicknesses of very thin plates of transparent bodies. Whence to know the causes of the changes of colours which are often made by the mixtures of several liquors, it is to be considered how the particles of any tincture may have their size or density altered by the infusion of another liquor.

When any metal is put into common water, the water cannot enter into its pores to act on it & dissolve it. Not that water consists of too gross parts for this purpose, but because it is unsociable to metal. For there is a certain secret principle in nature by which liquors are sociable to some things & unsociable to others. Thus water will not mix with oyle but readily with spirit of wine or with salts. It sinks also into wood which Quicksilver will not, but Quicksilvers sinks into metals, which, as I said, water will not. So Aqua fortis dissolves silver not gold; Aqua regis gold & not silver, &c. But a liquor which is of it self unsociable to a body may by the mixture of a convenient mediator be made sociable. So molten Lead which alone will not mix with copper or with Regulus of Mars, by the addition of Tin is made to mix with either. And water by the mediation of saline spirits will mix with metal. Now when any metal is put in water impregnated with such spirits, as into Aqua fortis, Aqua Regis, spirit of Vitriol or the like, the particles of the spirits as they in floting in the water, strike on the metal, will by their sociableness enter into its pores & gather round its outside particles, & by advantage of the continual tremor the particles of the metal are in, hitch themselves in by degrees between those particles & the body & loosen them from it, & the water entring into the pores together with the saline spirits, the particles of the metal will be thereby still more loosed, so as by that motion the solution puts them into, to be easily shaken of & made to Rote in the water: the saline particles still encompassing the metallick ones as a coat or shell does a kernell, after the manner expressed 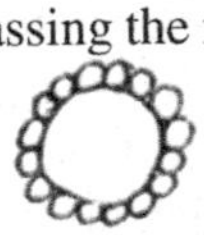 in the annexed figure. In which figure I have made the particles round, though they may be cubical or of any other shape.

If into a solution of metal thus made, be poured a liquor abounding with particles, to which the former saline particles are more sociable then to the particles of the metal, (suppose with particles of salt of Tartar:) then so soon as they strike on one another in the liquor, the saline particles will adhere to those more firmly then to the metalline ones, & by degrees be wrought of from those to enclose these. Suppose A a metalline particle enclosed with saline ones of spirit of Nitre, & E a

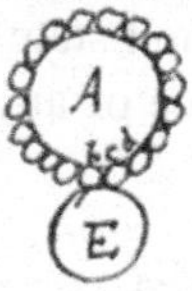

particle of salt of Tartar contiguous to two of the particles of spirit of nitre b & c, & suppose the particle E is impelled by any motion towards d so as to roll about the particle c till it touch the particle d: the particle b adhering more firmly to E then to A, will be forced off from A. And by the same means the particle E as it rolls about A will tear of the rest of the saline particles from A, one after another, till it has got them all or almost all about it self. And when the metallic particles are thus divested of the nitrous ones which as a mediator between them & the water held them floting in it: the Alcalizate ones crouding for the room the metallic ones took up before, will press these towards one another & make them come more easily together: so that by the motion they continually have in the water they shall be made to strike on one another, & then by means of the principle in the second supposition they will cohere & grow into clusters, & fall down by their weight to the bottom, which is called precipitation.

In the solution of metals, when a particle is loosing from the body, so soon as it gets to that distance from it where the principle of receding described in the 4th & 5th suppositions begins to overcome the principle of acceding described in the second supposition: the receding of the particle will be thereby accelerated, so that the particle shall as were with violence leap from the body, & putting the liquor into a brisk agitation, beget & promote that heat we often find to be caused in solutions of Metals. And if any particle happen to leap of thus from the body before it be surrounded with water, or to leap of with that smartness as to get loos from the water: the water by the principle in the 4th & 5th suppositions, will be kept of from the particle & stand round about it like a spherically hollow arch, not being able to come to a full contact with it any more. And severall of these particles afterwards gathering into a cluster, so as by the same principle to stand at a distance from one another without any water between them, will compose a buble. Whence I suppose it is that in brisk solutions there usually happens an ebullition.

This is one way of transmuting gross compact substances into aereal ones. Another way is by heat. For as fast as the motion of heat can shake off the particles of water from the surface of it: those particles by the said principle will Rote up & down in the air at a distance both from one another & from the particles of air, &make that substance we call vapor. Thus I suppose it is when the particles of a body are very small (as I suppose those of water are) so that the action of heat alone may be sufficient to shake them asunder. But if the particles be much larger, they then require the greater force of dissolving Menstruums to separate them, unless by any means the particles can be first broken into smaller ones. For the most fixed bodies, even Gold it self, some have said will become volatile only by breaking their parts smaller. Thus may the volatility & fixedness of bodies depend on the different sizes of their parts.

And on the same difference of size may depend the more or less permanency of aereal substances in their state of rarefaction. To understand this let us suppose ABCD to be a large piece of any metal, EFGH the limit of the interior uniform æther, & K

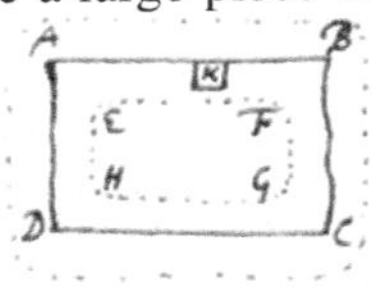

a part of the metal at the superficies AB. If this part or particle K be so little that it reaches not to the limit EF, its plain that the æther at its center must be less rare then if the particle were greater, for were it greater, its center would be further from the superficies AB, that is, in a place where the æther (by supposition) is rarer. The less the particle K therefore, the denser the æther at its center, because its center comes nearer to the edge AB where the æther is denser then within the limit EFGH. And if the particle were divided from the body & removed to a distance from it where the æther is still denser, the æther within it must proportionally grow denser. If you consider this you may apprehend how by diminishing the particle, the rarity of the æther within it will be diminished, till between the density of the æther without & the density of the æther within it there be little difference, that is till the cause be almost taken away which should keep this & other such particles at a distance from one another. For that cause, explained in the 4th & 5th suppositions, was the excess of density of the external æther above that of the internal. This may be the reason then why the small particles of vapors easily come together & are reduced back into water unless the heat which keeps them in agitation be so great as to dissipate them as fast as they come together: but the grosser particles of exhalations raised by fermentation keep their aerial form more obstinately, because the æther within them is rarer.

Nor does the size only but the density of the particles also conduce to the permanency of aereal substances. For the excess of density of the æther without such particles above that of the æther within them is still greater. Which has made me sometimes think that the true permanent Air may be of a metallic original: the particles of no substances being more dense then those of metals. This I think is also favoured by experience for I remember I once read in the Philosophical Transactions how M. Hugens at Paris found that the air made by dissolving salt of Tartar would in 2 or 3 days time condense & fall down again, but the air made by dissolving a metal continued without condensing or relenting in the least. If you consider then how by the continual fermentations made in the bowels of the earth there are aereal substances raised out of all kinds of bodies, all which together make the Atmosphere & that of all these the metallic are the most permanent, you will not perhaps think it absurd that the most permanent part of the Atmosphere, which is the true air, should be constituted of these: especially since they are the heaviest of all other & so must subside to the lower parts of the Atmosphere & float upon the surface of the earth, & buoy up the lighter exhalation & vapours to float in greatest plenty above them. Thus I say it ought to be with the metallic exhalations raised in the bowels of the earth by the action of acid menstruums, & thus it is with the true permanent air. For this as in reason it ought to be esteemed the most ponderous part of the Atmosphere because the lowest: so it betrays its ponderosity by making vapors ascend readily in it, by sustaining mists & clouds of snow, & by buoying up gross & ponderous smoke. The air also is the most gross unactive part of the Atmosphere affording living things no nourishment if deprived of the more tender exhalations & spirits that flote in it: & what more unactive & remote from nourishment then metallick bodies.

I shall set down one conjecture more which came into my mind now as I was writing this letter. It is about the cause of gravity. For this end I will suppose æther to consist of parts differing from one another in subtilty by indefinite degrees: That in the pores of bodies there is less of the grosser æther in proportion to the finer then in open spaces, & consequently that in the great body of the earth there is much less of the grosser æther in proportion to the finer then in the regions of the air: & that yet the grosser æther in the Air affects the upper regions of the earth & the finer æther in the earth the lower regions of the air, in such a manner that from the top of the air to the surface of the earth & again from the surface of the earth to the center thereof the æther is insensibly finer & finer. Imagin now any body suspended in the air or lying on the earth: & the æther being by the Hypothesis grosser in the pores which are in the upper parts of the body then in those which are in its lower parts, & that grosser æther being less apt to be lodged in those pores then the finer æther below, it will endeavour to get out & give way to the finer æther below, which cannot be without the bodies descending to make room above for it to go out into.

From this supposed gradual subtilty of the parts of æther some things above might be further illustrated & made more intelligible, but by what has been said you will easily discern whether in these conjectures there be any degree of probability, which is all I aim at. For my own part I have {so} little fansy to things of this nature that had not your encouragement moved me to it, I should never I think have thus far set pen to paper about them. What's amiss therefore I hope you will the more easily pardon in

Your most humble Servant & honourer
Is. Newton.
Cambridge Feb 28.
$167^{\frac{8}{9}}$
Source: The Newton Project
http://www.newtonproject.sussex.ac.uk/view/texts/normalized/NATP00275

Appendix 2

A Letter to a Physician

"Sir,—Not being acquainted with this handwriting, you will probably look to the bottom to find the subscription, and not finding any, will certainly wonder at this strange method of addressing to you. I must here in the beginning beg you to excuse it, and, to persuade you to read what follows with some attention, must tell you, that this gives you an opportunity to do a very good-natured action, which I believe is the most powerful argument I can use. I need not tell you, that I am your countryman, a Scotsman; for without any such tie, I dare rely upon your humanity even to a perfect stranger, such as I am. The favour I beg of you is your advice, and the reason why I address myself in particular to you need not be told,—as one must be a skilful physician, a man of letters, of wit, of good sense, and of great humanity, to give me a satisfying answer. I wish fame had pointed out to me more persons, in whom these qualities are united, in order to have kept me some time in suspense. This I say in the sincerity of my heart, and without any intention of making a compliment; for though it may seem necessary, that, in the beginning of so unusual a letter, I should say some fine things, to bespeak your good opinion, and remove any prejudices you may conceive at it, yet such an endeavour to be witty, would ill suit with the present condition of my mind; which, I must confess, is not without anxiety concerning the judgment you will form of me. Trusting, however, to your candour and generosity, I shall, without further preface, proceed to open up to you the present condition of my health, and to do that the more effectually, shall give you a kind of history of my life, after which you will easily learn why I keep my name a secret.

"You must know then that, from my earliest infancy, I found always a strong inclination to books and letters. As our college education in Scotland, extending little further than the languages, ends commonly when we are about 14 or 15 years of age, I was after that left to my own choice in my reading, and found it incline me almost equally to books of reasoning and philosophy, and to poetry and the polite authors. Every one who is acquainted either with the philosophers or critics, knows that there is nothing yet established in either of these two sciences, and that they contain little more than endless disputes, even in the most fundamental articles. Upon examination of these, I found a certain boldness of temper growing in me, which was not inclined to submit to any authority in these subjects, but led me to seek out some new medium, by which truth might be established. After much study and reflection on this, at last, when I was about 18 years of age, there seemed to be opened up to me a new scene of thought, which transported me beyond measure, and made me, with an ardour natural to young men, throw up every other pleasure or business to apply entirely to it. The law, which was the business I designed to follow, appeared nauseous to me, and I could think of no other way of pushing my fortune in the world, but that of a scholar and philosopher. I was infinitely happy in this course of life for some months; till at last, about the beginning of September, 1729, all my ardour seemed in a moment to be extinguished, and I could no longer raise my mind to that pitch, which formerly gave me such excessive pleasure. I felt no uneasiness or want of spirits, when I laid aside my book; and therefore never imagined there was any bodily distemper in the case, but that my coldness proceeded from a laziness of temper, which must be overcome by redoubling my application. In this condition I remained for 9 months, very uneasy to myself, as you may well imagine, but without growing any worse, which was a miracle. There was another particular, which contributed, more than any thing, to waste my spirits and bring on me this distemper, which was, that having read many books of morality, such as Cicero, Seneca, and Plutarch, and being smit with their beautiful representations of virtue and philosophy, I undertook the improvement of my temper and will, along with my reason and understanding. I was continually fortifying myself with reflections against death, and poverty, and shame, and pain, and all the other calamities of life. These no doubt are exceeding useful, when joined with an active life, because the occasion being presented along with the reflection, works it into the soul, and makes it take a deep impression; but in solitude they serve to little other purpose, than to waste the spirits, the force of the mind meeting with no resistance, but wasting itself in the air, like our arm when it misses its aim. This, however, I did not learn but by experience, and till I had already ruined my health, though I was not sensible of it. Some scurvy spots broke out on my fingers the first winter I fell ill, about which I consulted a very knowing physician, who gave me some medicine that removed these symptoms, and at the same time gave me a warning against the vapours, which, though I was labouring under at that time, I fancied myself so far removed from, and indeed from any other disease, except a slight scurvy, that I despised his warning. At last, about April 1730, when I was 19 years of age, a symptom, which I had noticed a little from the beginning, increased considerably; so that, though it was no uneasiness, the novelty of it made me ask advice; it was what they call a ptyalism or wateryness in the mouth. Upon my mentioning it to my physician, he laughed at me, and told me I was now a brother, for that I had fairly got the disease of the learned. Of this he found great difficulty to persuade me, finding in myself nothing of that lowness of spirit, which those who labour under that distemper so much complain of. However upon his advice I went under a course of bitters, and anti-hysteric pills, drank an English pint of claret wine every day, and rode 8 or 10 Scotch miles. This I continued for about 7 months after.

"Though I was sorry to find myself engaged with so tedious a distemper, yet the knowledge of it set me very much at ease, by satisfying me that my former coldness proceeded not from any defect of temper or genius, but from a disease to which any one may be subject. I now began to take some indulgence to myself; studied moderately, and only when I found my spirits at their highest pitch, leaving off before I was weary, and trifling away the rest of my time in the best manner I could. In this way, I lived with satisfaction enough; and on my return to town next winter found my spirits very much recruited, so that, though they sank under me in the higher flights of genius, yet I was able to make considerable progress in my former designs. I was very regular in my diet and way of life from the beginning, and all that winter made it a constant rule to ride twice or thrice a-week, and walk every day. For these reasons, I expected, when I returned to the country, and could renew my exercise with less interruption, that I would perfectly recover. But in this I was much mistaken; for next summer, about May 1731, there grew upon me a very ravenous appetite, and as quick a digestion, which I at first took for a good symptom, and was very much surprised to find it bring back a palpitation of heart, which I had felt very little of before. This appetite, however, had an effect very unusual, which was to nourish me extremely; so that in 6 weeks' time, I passed from the one extreme to the other; and being before tall, lean, and raw-boned, became on a sudden the most sturdy, robust, healthful-like fellow you have seen, with a ruddy complexion and a cheerful countenance. In excuse for my riding, and care of my health, I always said that I was afraid of consumption, which was readily believed from my looks, but now every body congratulated me upon my thorough recovery. This unnatural appetite wore off by degrees, but left me as a legacy the same palpitation of the heart in a small degree, and a good deal of wind in my stomach, which comes away easily, and without any bad *goût*, as is ordinary. However, these symptoms are little or no uneasiness to me. I eat well; I sleep well; have no lowness of spirits, at least never more than what one of the best health may feel from too full a meal, from sitting too near a fire, and even that degree I feel very seldom, and never almost in the morning or forenoon. Those who live in the same family with me, and see me at all times, cannot observe the least alteration in my humour, and rather think me a better companion than I was before, as choosing to pass more of my time with them. This gave me such hopes, that I scarce ever missed a day's riding, except in the winter time; and last summer undertook a very laborious task, which was to travel eight miles every morning, and as many in the forenoon, to and from a mineral well of some reputation. I renewed the bitter and anti-hysteric pills twice, along with anti-scorbutic juice, last spring, but without any considerable effect, except abating the symptoms for a little time.

"Thus I have given you a full account of the condition of my body; and without staying to ask pardon, as I ought to do, for so tedious a story, shall explain to you how my mind stood all this time, which on every occasion, especially in this distemper, have a very near connexion together. Having now time and leisure to cool my inflamed imagination, I began to consider seriously how I should proceed in my philosophical inquiries. I found that the moral philosophy transmitted to us by antiquity laboured under the same inconvenience that has been found in their natural philosophy, of being entirely hypothetical, and depending more upon invention than experience: every one consulted his fancy in erecting schemes of virtue and of happiness, without regarding human nature, upon which every moral conclusion must depend. This, therefore, I resolved to make my principal study, and the source from which I would derive every truth in criticism as well as morality. I believe it is a certain fact, that most of the philosophers who have gone before us, have been overthrown by the greatness of their genius, and that little more is required to make a man succeed in this study, than to throw off all prejudices either for his own opinions or for those of others. At least this is all I have to depend on for the truth of my reasonings, which I have multiplied to such a degree, that within these 3 years, I find I have scribbled many a quire of paper, in which there is nothing contained but my own inventions. This, with the reading most of the celebrated books in Latin, French, and English, and acquiring the Italian, you may think a sufficient business for one in perfect health, and so it would had it been done to any purpose; but my disease was a cruel encumbrance on me. I found that I was not able to follow out any train of thought, by one continued stretch of view, but by repeated interruptions, and by refreshing my eye from time to time upon other objects. Yet with this inconvenience I have collected the rude materials for many volumes; but in reducing these to words, when one must bring the idea he comprehended in gross, nearer to him, so as to contemplate its minutest parts, and keep it steadily in his eye, so as to copy these parts in order,—this I found impracticable for me, nor were my spirits equal to so severe an employment. Here lay my greatest calamity. I had no hopes of delivering my opinions with such elegance and neatness, as to draw to me the attention of the world, and I would rather live and die in obscurity than produce them maimed and imperfect.

"Such a miserable disappointment I scarce ever remember to have heard of. The small distance betwixt me and perfect health makes me the more uneasy in my present situation. It is a weakness rather than a lowness of spirits which troubles me, and there seems to be as great a difference betwixt my distemper and common vapours, as betwixt vapours and madness. I have noticed in the writings of the French mystics, and in those of our fanatics here, that when they give a history of the situation of their souls, they mention a coldness and desertion of the spirit, which frequently returns; and some of them, at the beginning, have been tormented with it many years. As this kind of devotion depends entirely on the force of passion, and consequently of the animal spirits, I have often thought that their case and mine were pretty parallel, and that their rapturous admirations might discompose the fabric of the nerves and brain, as much as profound reflections, and that warmth or enthusiasm which is inseparable from them.

"However this may be, I have not come out of the cloud so well as they commonly tell us they have done, or rather began to despair of ever recovering. To keep myself from being melancholy on so dismal a prospect, my only security was in peevish reflections on the vanity of the world and of all human glory; which, however just sentiments they may be esteemed, I have found can never be sincere, except in those who are possessed of them. Being sensible that all my philosophy would never make me contented in my present situation, I began to rouse up myself; and being encouraged by instances of recovery from worse degrees of this distemper, as well as by the assurances of my physicians, I began to think of something more effectual than I had hitherto tried. I found, that as there are two things very bad for this distemper, study and idleness, so there are two things very good, business and diversion; and that my whole time was spent betwixt the bad, with little or no share of the good. For this reason I resolved to seek out a more active life, and though I could not quit my pretensions in learning but with my last breath, to lay them aside for some time, in order the more effectually to resume them. Upon examination, I found my choice confined to two kinds of life, that of a travelling governor, and that of a merchant. The first, besides that it is in some respects an idle life, was, I found, unfit for me; and that because from a sedentary and retired way of living, from a bashful temper, and from a narrow fortune, I had been little accustomed to general companies, and had not confidence and knowledge enough of the world to push my fortune, or to be serviceable in that way. I therefore fixed my choice upon a merchant; and having got recommendation to a considerable trader in Bristol, I am just now hastening thither, with a resolution to forget myself, and every thing that is past, to engage myself, as far as is possible, in that course of life, and to toss about the world, from the one pole to the other, till I leave this distemper behind me.

"As I am come to London in my way to Bristol, I have resolved, if possible, to get your advice, though I should take this absurd method of procuring it. All the physicians I have consulted, though very able, could never enter into my distemper; because not being persons of great learning beyond their own profession, they were unacquainted with these motions of the mind. Your fame pointed you out as the properest person to resolve my doubts, and I was determined to have somebody's opinion, which I could rest upon in all the varieties of fears and hopes, incident to so lingering a distemper. I hope I have been particular enough in describing the symptoms to allow you to form a judgment; or rather, perhaps, have been too particular. But you know it is a symptom of this distemper, to delight in complaining and talking of itself. The questions I would humbly propose to you are: Whether, among all those scholars you have been acquainted with, you have ever known any affected in this manner? Whether I can ever hope for a recovery? Whether I must long wait for it? Whether my recovery will ever be perfect, and my spirits regain their former spring and vigour, so as to endure the fatigue of deep and abstruse thinking? Whether I have taken a right way to recover? I believe all proper medicines have been used, and therefore I need mention nothing of them."

March/April 1734
Source:
LIFE AND CORRESPONDENCE OF DAVID HUME.
FROM THE PAPERS BEQUEATHED BY HIS NEPHEW TO THE ROYAL SOCIETY OF EDINBURGH; AND OTHER ORIGINAL SOURCES.
By JOHN HILL BURTON, Esq. ADVOCATE.
VOLUME I. EDINBURGH: WILLIAM TAIT, 107, PRINCE'S STREET.
MDCCCXLVI.

Index

Z. Parusniková, *David Hume, Sceptic*, SpringerBriefs in Philosophy,
DOI 10.1007/978-3-319-43794-1

The manufacturer's authorised representative in the EU is Springer Nature Customer Service Centre GmbH, Europaplatz 3, 69115 Heidelberg, Germany. If you have any concerns regarding our products, please contact ProductSafety@springernature.com

Printed and bound by CPI Group (UK) Ltd, Croydon, CR0 4YY
22/09/2026
02221064-0018